CAMBRIDGE

Book:

M000189761

Life Sciences

Until the nineteenth century, the various subjects now known as the life sciences were regarded either as arcane studies which had little impact on ordinary daily life, or as a genteel hobby for the leisured classes. The increasing academic rigour and systematisation brought to the study of botany, zoology and other disciplines, and their adoption in university curricula, are reflected in the books reissued in this series.

On the Growth of Plants in Closely Glazed Cases

In the early nineteenth century, live plant cuttings were commonly transported between continents in wooden boxes exposed to the elements on the decks of ships; unsurprisingly, it was rare for them to arrive in good health. The glass cases devised by Nathaniel Bagshaw Ward (1791–1868) were a revolutionary step forward in preserving botanical specimens. In this monograph, first published in 1842, Ward explores some of the most common causes of plant deaths in cities and aboard ships, including air quality and temperature. Most importantly, he emphasises the need for light. Although photosynthesis would not be chemically understood until later that century, Ward recognised that a glass case was infinitely preferable to an opaque one. His rapidly adopted invention would have far-reaching effects, allowing for the safe transportation of tea from China to the Himalayas, rubber from the Amazon and medicinal species from the Andes to India.

Cambridge University Press has long been a pioneer in the reissuing of out-of-print titles from its own backlist, producing digital reprints of books that are still sought after by scholars and students but could not be reprinted economically using traditional technology. The Cambridge Library Collection extends this activity to a wider range of books which are still of importance to researchers and professionals, either for the source material they contain, or as landmarks in the history of their academic discipline.

Drawing from the world-renowned collections in the Cambridge University Library and other partner libraries, and guided by the advice of experts in each subject area, Cambridge University Press is using state-of-the-art scanning machines in its own Printing House to capture the content of each book selected for inclusion. The files are processed to give a consistently clear, crisp image, and the books finished to the high quality standard for which the Press is recognised around the world. The latest print-on-demand technology ensures that the books will remain available indefinitely, and that orders for single or multiple copies can quickly be supplied.

The Cambridge Library Collection brings back to life books of enduring scholarly value (including out-of-copyright works originally issued by other publishers) across a wide range of disciplines in the humanities and social sciences and in science and technology.

On the Growth of Plants in Closely Glazed Cases

NATHANIEL BAGSHAW WARD

CAMBRIDGE
UNIVERSITY PRESS

CAMBRIDGE UNIVERSITY PRESS

Cambridge, New York, Melbourne, Madrid, Cape Town,
Singapore, São Paolo, Delhi, Mexico City

Published in the United States of America by Cambridge University Press, New York

www.cambridge.org
Information on this title: www.cambridge.org/9781108061131

© in this compilation Cambridge University Press 2013

This edition first published 1842
This digitally printed version 2013

ISBN 978-1-108-06113-1 Paperback

ON THE

GROWTH OF PLANTS

IN

CLOSELY GLAZED CASES.

ON THE

GROWTH OF PLANTS

IN

CLOSELY GLAZED CASES:

BY

N. B. WARD, F.L.S.

WISDOM OF GOD IN CREATION.

LONDON :

JOHN VAN VOORST, PATERNOSTER ROW.

M.DCCC.XLII.

Μηδὲν εἴσετο κακὸν.

PREFACE.

Most of the facts detailed in the following work have been long before the public. They have been published in several periodicals, and in a letter to Sir W. J. Hooker, which appeared in the 'Companion to the Botanical Magazine,' for May, 1836. This letter was subsequently printed separately for private distribution among the author's friends. The attention of the scientific world was likewise drawn to the subject at three several meetings of the British Association, and more particularly by an admirable paper written by the late Mr. Ellis of Edinburgh, and published in the 'Gardener's Magazine' for September, 1839. The simple yet comprehensive principle however upon which plants are grown in closed cases, does not appear to be clearly understood, and many misconceptions yet exist upon this point. The object of the present work therefore is to remove these erroneous notions, and thereby to enable those who wish to experiment upon the subject to do so without risk of disappointment. The author is fearful that in this

attempt he will be condemned by the learned for
having entered into needless details, while to those
who are wholly unacquainted with the leading princi-
ples of botanical science he may not have rendered
his meaning sufficiently clear. He has, however,
done his best, and can only say in the oft-quoted
words of the poet —

> " si quid novisti rectius istis
> Candidus imperti, si non,—his utere mecum."

A grateful task remains. To the Messrs. Loddiges,
who may most justly be styled 'Hortulanorum Prin-
cipes,' the thanks of the author are most especially
due. From the very commencement of his enquiries
their splendid stores were placed unreservedly at his
disposal, and without their kind assistance it would
have been difficult for him to have carried on his experiments. He begs likewise to record his obligations
to Mr. Aiton, Sir W. J. Hooker, and Mr. Smith,
of Kew; to his old friend Mr. Anderson, of the
Chelsea Botanic Garden; to Dr. Lindley, of the
Horticultural Society's Garden; to Mr. Macnab of
Edinburgh; Mr. Mackay of Dublin; Mr. Cameron of
Birmingham; and to various other friends, for numerous specimens of interesting plants received from
them at different times for the purposes of experiment.

Wellclose Square,
March, 1842.

TABLE OF CONTENTS.

——

PAGE.

CHAP. I.—On the Natural Conditions of Plants.   -   3

CHAP. II.—On the causes which interfere with the Natural Conditions of Plants in large Towns, &c.   -   11

CHAP. III.—On the Imitation of the Natural Conditions of Plants
in closely-glazed Cases.   -   -   25

CHAP. IV.—On the conveyance of Plants and Seeds on Ship-board.   45

CHAP. V.—On the application of the closed plan in improving the
condition of the poor.   -   57

CHAP. VI.—On the probable future application of the preceding facts.   67

Appendix.   -   -   75

CHAPTER 1.

ON THE NATURAL CONDITIONS OF PLANTS.

" There, fed by food they ove, to rankest size,
  Around the dwelling docks and wormwood rise;
  Here the strong mallow strikes her slimy root;
  Here the dull nightshade hangs her deadly fruit;
  On hills of dust the henbane's faded green,
  And pencilled flower of sickly scent is seen;
  At the wall's base the fiery nettle springs,
  With fruit globose, and fierce with poisoned stings.
  Above, the growth of many a year is spread,
  The yellow level of the stonecrop's bed;
  In every chink delights the fern to grow,
  With glossy leaf and tawny bloom below."

CRABBE.

CHAPTER I.

ON THE NATURAL CONDITIONS OF PLANTS.

To enter into any lengthened detail on the all-important subject of the natural conditions of plants would occupy far too much space; yet to pass it by without special notice, in any work treating of their cultivation, would be impossible.

Without a knowledge of the laws which regulate their growth, all our attempts must be empirical and more or less abortive. If we examine the vegetation on the surface of the globe, we shall find that the circumstances under which plants exist and flourish vary in an endless degree, and that they are all influenced by the atmosphere, heat, light, moisture, varieties of soil and periods of rest.

The purity of the atmosphere most sensibly affects the growth of plants, as evinced in the difference between those which grow in London and other large towns, or within the reach of manufactories evolving noxious gases, and those which grow in the country; but of this more hereafter.

The heat to which plants are subjected varies from 32° to 170°, or 180°. Thus, in some parts of Mexico, the heat is so intense and the soil and atmosphere so dry, that no vegetation is found at certain seasons, save a species of *Cactus*, and were it not for this plant these tracts would form impassable barriers.   Hardy states, in his Travels, that the sole subsistence of himself and party for four days consisted of the fruit of the *Petaya*, which, unlike most other luscious fruits, rather removes than creates thirst, while, at the same time, it satisfies to a certain degree the sensation of hunger.   The providence of God is equally manifested in cold countries, as in Lapland, where the rein-deer moss furnishes the sole food, during winter, for the rein-deer, without which the inhabitants could not exist.

It is hardly possible to overrate the influence of light upon plants ; but its intensity varies from almost total darkness to a light double that of our brightest summer's day.   Upon light depend all the active properties, the color, &c.   I am tempted to give an example from Mr. Ellis, of its effects in this last particular :—" In North America the operation of light in colouring the leaves of plants is sometimes exhibited on a great scale, and in a very striking manner.   Over the vast forests of that country clouds sometimes spread, and continue for many days, so as almost entirely to intercept the rays of the sun.   In one instance, just about the period of vernation, the sun had not shone for twenty days, during which time the leaves of the trees had reached nearly their full size, but were of a pale or whitish color.   One forenoon the sun broke through in full brightness, and the color of the leaves changed so fast, that, by the middle of the afternoon, the whole forest for many miles in length exhibited its usual summer's dress."

The states of moisture vary as much as those of heat and light. The late Mr. Allan Cunningham often expressed to me his surprise at the extreme dryness of the atmosphere and soil in which grew many species of New Holland plants, — that in seasons when there was neither dew nor rain, he had dug several feet below their roots without finding a trace of water, and yet Banksias and Acacias would continue to live in this state for a considerable time. There are numerous other plants, independently of those which live in water, which cannot exist unless the atmosphere and soil are continually humid, such as *Trichomanes speciosum*, &c. &c. Plants are affected by soils — sometimes specifically — but more generally in consequence of various soils possessing different powers of imbibing and retaining moisture. All plants require rest, and obtain it in some countries by the rigour of winter; in others by the scorching and arid heat of summer. Some " after short slumber wake to life again," while the sleep of others is unbroken for many months. This is the case with most alpine plants, and is necessary to their well-being. Messrs. Balfour and Babington, whilst recently exploring the lofty mountains of Harris, found the climate to be so modified by the vicinity of the Great Atlantic Ocean, that, notwithstanding their northern latitude (58°), many of the species inhabiting the Highland districts of Scotland were wholly wanting, and that the few which they saw were confined to the coldest and most exposed spots. From the same cause many plants grow there which were not known to grow in so northern a latitude in Britain. In Egypt the blue water-lily obtains rest in a curious way. Mr. Traill, the gardener of Ibrahim Pacha, informed me that this plant abounds in several of the canals at Alexandria,

which at certain seasons become dry; and the beds of
these canals, which quickly become burnt as hard as bricks
by the action of the sun, are then used as carriage-roads.
When the water is again admitted the plant resumes its
growth with redoubled vigour.

To suit all the varied conditions to which I have thus
briefly alluded, and under which plants are found to exist,
they have been formed by their Almighty Creator of different structures and constitutions, to fit them for the
stations they severally hold in creation; and so striking
are the results, that every different region of the globe is
characterized by peculiar forms of vegetation. A practised
botanical eye can with certainty, in almost all cases, predict
the capabilities of any hitherto unknown country, by an inspection of the plants which it produces.   It were much
to be wished, that those upon whom the welfare of thousands of their starving emigrant countrymen depends,
possessed a little more of this most useful knowledge.
But in order to give us a clearer idea of the " strong con-
nexions, nice dependencies," existing between climate
and vegetation, let us survey plants in a state of nature.
We shall find some restricted to certain situations, while
others have a wide range, or greater powers of adaptation.   It is not perhaps going too far to assert, that
no two plants are alike in this particular, or in other
words, that the constitution of every individual plant is
different.   Of the former, *Trichomanes speciosum* is an
example, it not being able to exist, even for a short time,
in a dry atmosphere : of the latter, familiar examples are
presented to us in the London Pride and the Auricula;
these of course grow in greater or less luxuriance, as the
conditions are more or less favorable. The *Cerasus Virgi-*

niana affords an interesting illustration of the effects of climate upon vegetation : in the southern states of America it is a noble tree, attaining one hundred feet in height; in the sandy plains of the Saskatchawan it does not exceed twenty feet; and at its northern limit, the Great Slave Lake, in lat. 62°, it is reduced to a shrub of five feet. But we need not travel to America to seek instances of this sort: we have them every where about us. I have gathered, on the chalky borders of a wood in Kent, perfect specimens in full flower of *Erythræa Centaurium*, not more than half an inch in height, consisting of one or two pairs of most minute leaves, with one solitary flower: these were growing on the bare chalk. By tracing the plant towards and in the wood, I found it gradually increasing in size, until its full development was attained in the open parts of the wood, where it became a glorious plant, four or five feet in elevation, and covered with hundreds of flowers. Let us pause here a moment, and reflect deeply on the wonders around us. We shall find a continued succession of beauties throughout the year, beginning with the Primrose, the Violet, and the Anemone; these giving place to the Orchises; and these again to the Mulleins, Campanulas, and various other plants, all in their turn delighting the eye and gladdening the heart: nor is the winter season devoid of interest; the surface of the ground, and every decaying leaf and twig, is inhabited by a world of microscopic beauties. All these have maintained their ground, without interfering with each other, year after year and generation after generation. The same page in the great book of nature, which filled the mind of Ray with the Wisdom of God in Creation, lies open to our view. May we read it aright! Let us ask ourselves, whether man, with all his boasted wisdom, can realize such a scene as this ?   He cannot,—and the cause lies in his ignorance of the natural conditions of plants.   To sum up in the words of a great philosopher of the present day.— " If the laws of nature, on the one hand, are invincible opponents, on the other they are irresistible auxiliaries ; and it will not be amiss if we regard them in each of these characters, and consider the great importance of them to mankind.

" 1stly. In showing us how to avoid attempting impossibilities.

" 2dly. In securing us from important mistakes in attempting what is in itself possible, by means either inadequate, or actually opposed to the ends in view.

" 3dly. In enabling us to accomplish our ends in the easiest, shortest, most economical and most effectual manner.

"4thly. In inducing us to attempt, and enabling us to accomplish objects, which, but for such knowledge, we should never have thought of undertaking."—*Herschel*.

CHAPTER II.

ON THE CAUSES WHICH INTERFERE WITH THE NATURAL CONDITIONS OF PLANTS IN LARGE TOWNS.

"As well might corn, as verse, in cities grow ;
In vain the thankless glebe we plough and sow :
Against th' unnatural soil in vain we strive ;
'T is not a ground in which these plants will thrive."

COWLEY.

CHAPTER II.

AMONG the causes tending to depress vegetation in large towns may be mentioned *deficiency of light,* the *dryness of the atmosphere,* the *fuliginous matter* with which the air of large towns is always more or less loaded, and the *evolution of noxious gases* from manufactories.

When we consider the all-important agency of *light* in the functions of the vegetable system, that upon it depend the nutrition of the plant, the formation of its secretions, &c., we shall not err when we attribute a portion of the depressing effects upon some plants to *deficiency of light;* but that this cannot be the sole cause is clear from the impossibility of growing such plants as ferns and mosses, which can, in any part of London, obtain as much light as they require.

With respect to the *dryness of the atmosphere,* my friend Mr. W. H. Campbell, late secretary to the Edinburgh Botanical Society, thus writes. — " It occurs to me, that the *want of moisture* in a town-atmosphere is the greatest enemy with which vegetation has to contend; and it seems obvious, that the larger the space occupied by a town, the greater must this want of moisture be. Stone houses, walls, and streets, are all the ready absorbents, or reflectors of heat; and whatever rain falls is speedily drained off their surface, and carried away far from the town. Of course, the more fires, manufactories, and steam-engines there are, the dryness of the atmosphere will be the greater, and the power of vegetation be correspondingly reduced. Away from towns we find these circumstances exactly reversed. When rain falls the soil parts with no moisture until it can absorb no more. What is superfluous is then received in pools, ditches, marshes, lakes and streams, from all of which, and from the saturated soil, are exhaled those refreshing dews and vapours which so rarely visit the sickly vegetation of the town. In Edinburgh we have some instances of the power of moisture in obviating the disadvantages of proximity of smoke from manufactories. In some large pieces of garden-ground attached to houses at the south back of the Canongate, near Salisbury Craigs, it is almost impossible to rear herbaceous plants, or to preserve through a sickly existence any but the hardiest trees and shrubs, on account of their proximity to the heat and smoke of glass-works, breweries, &c. &c.; whereas, at the distance of 20 or 30 feet from these gardens, there is an extensive irrigated meadow, which, kept constantly moist, produces (despite of smoke and every other detrimental agent) ten or twelve crops in a season of the most luxuriant vegetation, the exhalations and irrigation in the driest summer constantly affording the requisite supply of moisture.

" In the same manner are we to account for the oases

in the deserts of Africa. Under the most unfavourable*
circumstances, in the midst of arid sand, Nature supplies a
spring of water, and immediately a luxuriant vegetation
ensues as far as its influence extends. So, in Wellclose
Square, I have no doubt, if a perennial fountain could be
made to meander through your court-yard, and sun and
light were freely admitted, that vegetation would even
there assume some of its loveliest forms, despite of smoke
and the other concomitants of a Life in London."

I believe, with Mr. Campbell, that a constant supply of
moisture would materially benefit vegetation in large
towns, but, from the result of my own experiments, and
from what every one may see in London itself, I cannot
imagine that " the dryness of the atmosphere is the greatest enemy to which the vegetation in London is subjected." Long before I contemplated the growth of plants
in closed cases I made an unsuccessful attempt to cultivate ferns† and various other plants in the open air, as
detailed at the commencement of the ensuing chapter.
If we examine old walls in London, which, from the
leakage of cisterns or pipes above them, are constantly
moist, we shall find a vegetation certainly, but not of a
healthy kind. The conditions for mosses will be so far
fulfilled as to allow of the development of their leaves ;
and we shall everywhere see, on such walls, the silvery

* Mr. C. is here under a mistake ; light and heat are abundant, and
moisture *alone* is wanting.

† Those who are desirous of acquiring an intimate acquaintance
with our native species of this beautiful and interesting order of plants,
and wish to cultivate them in the open air, cannot do better than consult
Mr. Newman's ' History of British Ferns.'

tips (when not obscured by soot) of *Bryum argenteum ;*
but we must go two or three miles out of London before
we can find it in fructification.   It is true that we may,
even in London, find the *Funaria hygrometrica* * in fruit,
but this is an exception to the general rule.   It is likewise
well known that, *cæteris paribus,* plants with smooth
leaves suffer less in London than those which are hairy, or
covered with viscid or resinous secretions.   Hence the
miserable appearance of most of the *Coniferæ* in London,
although these are plants, many species of which flourish
in the driest sands.

We must therefore look for another and more efficient
cause of depression, and this I believe is to be found
in the sooty particles diffused through the air.   In my
letter to Sir W. J. Hooker, published in the 'Companion
to the Botanical Magazine' for May, 1836, I expressed
my opinion that the depressing influence of the air of
large towns upon vegetation, depends *almost entirely*
upon the fuliginous matter with which such an atmo-

---

* The *Funaria hygrometrica* is a remarkable moss, differing widely in its
powers of adaptation, and consequently its greater geographical range,
from its congeners.   It appears to delight as much in heat as other
mosses do in cold.   There is nothing in its structure to lead us to infer
such a difference of constitution.   Most mosses are restricted within
certain limits, and will not fructify but under certain conditions.   The
Funaria is found in fruit not only in London, but in every brick-field
around it, — in my own fern-houses, — and likewise in Loddiges' Orchideous house, where the temperature often rises to 120°; and I have
specimens in my herbarium from all parts of the world; from Egypt,
the Cape of Good Hope, the East and West Indies, New Zealand,
New Holland, &c.   The peristome of this moss is one of our most
beautiful microscopic objects.

sphere is impregnated, and which produces similar effects upon the leaves of plants as upon the lungs of animals. This opinion has been questioned by the late Mr. Ellis, in an admirable paper published in the 'Gardener's Magazine' for September, 1839 ; and, as the subject is one of great importance — it being impossible to apply remedies without knowing the nature of the disease — I shall discuss it at some length. Mr. Ellis says that "the real mode in which such an atmosphere proves injurious to vegetation, was first shown by the experiments of Drs. Turner and Christison, which were published in the 93rd number of the 'Edinburgh Medical and Surgical Journal.' They ascertained that it is not simply to the diffusion of fuliginous matter through the air, but to the presence of sulphurous acid gas, generated in the combustion of coal, that the mischief is to be ascribed. When added to common air in the proportion of $\frac{1}{9000}$ or $\frac{1}{10000}$ part, that gas sensibly affected the leaves of growing plants in ten or twelve hours, and killed them in forty-eight hours or less. The effects of hydrochloric or muriatic acid gas were still more powerful, it being found that the tenth part of a cubic inch, in 20,000 volumes of air, manifested its action in a few hours, and entirely destroyed the plant in two days. Both these gases acted on the leaves, affecting more or less their color, and withering or crisping their texture, so that a gentle touch caused their separation from the footstalk; and both exerted this injurious operation, when present in such minute proportions as to be wholly inappreciable by the animal senses.

"After having suffered much injury from these acid gases, the plants, if removed in time, will recover, but

with the loss of their leaves. Hence, in vegetation carried
on in smoky atmospheres, the plants are rarely killed altogether, but merely blighted for a season. Accordingly, in
spring, vegetation recommences with its accustomed luxuriance, and as in many situations there is, at that season
and during the summer, a considerable diminution in the
number of coal fires, there will be a proportionate decrease
in the production of sulphurous acid gas; and, consequently, less injury will be done to plants during that
season. In winter, too, when coal fires mostly abound
and gas is most abundantly generated, deciduous plants
are protected from its noxious operation by suspension of
their vegetating powers ; but the leaves of evergreens,
which continue to grow through that season, are constantly
exposed to its action when present in its greatest intensity.
Accordingly, in many of the suburban districts round
London, especially in the course of the river where new
manufactories are constantly rising up, the atmosphere is
so highly charged with noxious matters, that many deciduous plants and almost all evergreens cease to flourish,
or exhibit only a sickly vegetation.

   "In an interesting biographical sketch of his late la-
mented friend Dr. Turner, Professor Christison confirms,
by subsequent experience, the opinion formerly given respecting the noxious operation of the sulphurous or muriatic acid gas upon plants. He describes their action as so
energetic, that, in the course of two days, the whole vegetation of various species of plants may be destroyed by
quantities so minute as to be altogether inappreciable by
the senses. On two occasions he was able to trace the
identical effects of the same kind of works (the black ash
manufactory) on the great scale, which his friend and

himself witnessed in their researches. In one instance the devastation committed was enormous, vegetation being for the most part miserably stunted, or blasted altogether, to a distance of fully a third of a mile from the works, in the prevailing direction of the wind. Against the evils arising from such a vitiated atmosphere, the plan of Mr. Ward provides effectual protection, as the success of his own establishment amply demonstrates."

The correctness of the above observations of Messrs. Turner and Christison, as to the effects of sulphurous and muriatic acid gases upon plants, cannot for one moment be doubted; and that plants suffer when exposed to a direct current of these gases, before there is time for diffusion through surrounding space, is equally matter of fact; but I contend, that it yet remains to be proved that there exists generally, in the atmosphere of London or other large cities, such a proportion of these noxious gases as sensibly to affect vegetation. We shall find in the windows of shops and small houses, in numerous parts of London, hundreds of geraniums and other plants, growing very well and without any crisping or curling of their leaves, care being taken in these instances to keep the plants perfectly clean, and free from soot; and it is certain, that although my cases can and do exclude the fuliginous portion of the atmosphere, and certainly protect the plants from the effects of any direct current of hurtful airs, they cannot exclude that portion which becomes mixed with the atmosphere. This subject is one of the highest importance, and intimately connected with the well-being of everything that hath life, whether vegetable or animal. It cannot, however, be understood, without reference to, and a full explanation of, the law which regulates the dif-

c

fusion of gases, — a law constantly in action under all
circumstances, and without the beneficent operation of
which, vegetable as well as animal life would suffer greatly
in large towns, and a cellar in St. Giles's quickly become
a *Grotto del Cane.*

" If we take two vessels, and fill one with carbonic acid
gas, and the other with hydrogen, (their weights respec-
tively being as 22 to 1), and then place the light gas
perpendicularly over the other, effecting a communication
between the vessels by means of a tube not larger in
diameter than a human hair, the two gases will immediately
begin to mix, and after a short interval will be found
equally distributed between both vessels. If the upper
vessel be filled with oxygen, nitrogen, or any other gas,
the same phenomena will ensue; the gases will be found,
after a short time, to be in a state of mixture, and at last
there will be equal portions of each in both vessels. The
permeability of animal membranes by gases has been fully
proved by the researches of Drs. Faust and Mitchell. It
fully appears from their experiments that animal membranes, both in the living and dead subject, both in and
out of the body, are freely penetrated by gaseous matter;
that the phenomena of endosmose and exosmose, observed
in liquids by Dutrochet, are likewise exhibited by gases.
If a glass full of carbonic acid be closed by an animal
membrane, or sheet of caoutchouc, and be then exposed
to the atmosphere, a portion of air will pass into the glass
and some of the confined gas escape from it; and if the
experiment be reversed, by confining air in the glass,
which is then placed in an atmosphere of carbonic acid,
the latter passes in and the former out of the glass. Similar phenomena ensue with other gases; so that when any

two gases are separated by a membrane, both of them pass through the partition. But though all gases pass through membranous septa, they differ remarkably in the relative rapidity of transmission. Thus, while a volume of carbonic acid requires $5\frac{1}{2}$ minutes to pass through a membrane, the same volume of oxygen requires 113, and a much greater time is required for nitrogen. Hence, when a bladder full of air is surrounded by carbonic acid, the latter enters faster than the former escapes, and the bladder bursts ; but on reversing the conditions of the experiment the bladder becomes flaccid, because the carbonic acid within passes out more rapidly than the exterior air enters. The transmission of gases in some of these experiments takes place in opposition to a pressure equal to several atmospheres."*

In order to ascertain what the degree of circulation through the substance of the membrane in some closed jars might have been, Dr. Daubeny† removed from one of the jars the plants and vegetable mould it contained, and then substituted for them about an equal amount of dry sand. He next passed through the vessel a current of oxygen, until the volume of air within contained no less than 77 per cent. of that gas. The air was examined again, at 4 p.m., after an interval of three hours from the period of the first experiment, when it was found to have lost 4 per cent. of oxygen. At 8 the next morning it was found to contain only 63 per cent. of oxygen, having diminished, in 16 hours, 10 per cent. After having been exposed all day to air and light, and examined at 8 p.m.,

* Turner's Elements.

† Dr. Daubeny's Report to the Meeting of the British Association, at Liverpool.

the oxygen was found to amount to only 45 per cent., having diminished, in 12 hours, 18 per cent. During the next night it had diminished, in 12 hours, only $6\frac{1}{2}$ per cent., the amount of oxygen the next morning being $38\frac{1}{2}$ per cent. During the next day it had lost 7 per cent., containing at 8 p.m. $31\frac{1}{2}$ per cent. The next night the diminution was only $2\frac{1}{2}$ per cent., and on the succeeding day 3 per cent. The following night the diminution was $1\frac{1}{2}$ per cent., the amount of oxygen being $24\frac{1}{2}$ per cent. only. During the day a further diminution of $3\frac{1}{2}$ per cent. took place, the air enclosed within the jar being found to contain exactly the quantity of oxygen present in atmospheric air. The following is a tabular view of the results.

June 23.	1 p.m. .... amount of oxygen, 77	... excess, 56
„ „	4 p.m. 73	 52
„ 24.	8 a.m. 63	 42
„ „	8 p.m. 45	 24
„ 25.	8 a.m. .... 38·5	.... 17·5
„ „	8 p.m. .... 31·5	.... 10·5
„ 26.	8 a.m. 29	 8
·, „	8 p.m. 26	 5
„ 27.	8 a.m. .... 24·5	.... 3·5
·, „	8 p.m. 21	.... 0·0

Thus five days were required to enable the whole excess of oxygen to pass through the substance of the membrane, the diameter of which was three inches, whilst the capacity of the vessel, when the sand had been introduced, was nearly one gallon, so that about three quarts of oxygen and one of nitrogen may be calculated as having been present in the jar at the commencement of the experiment, of which about $4\frac{1}{2}$ pints were discharged through the membrane in the course of the five days during which the observations were continued.

The transmission took place more rapidly during the day, because of the exposure of the jar to the sun and wind, which, by the expansion caused within the vessel, and by the more rapid succession of aërial currents brought into contact with the external surface of the membrane, doubtless caused in a greater degree the transmission of the redundant oxygen. The average quantity that escaped per diem did not much exceed 11 per cent., or did not quite amount to one pint in the 24 hours; but of course the transmission was more rapid at first, and diminished gradually in quantity, as the condition of the air within the jar approached more nearly to that of the atmosphere surrounding it.

"To conclude this curious subject, Spallanzani proved that some animals possessed of lungs, — such as serpents, lizards and frogs, — produce the same changes on the air by means of their skin, as by their proper respiratory organs; and Dr. Edwards, in a series of masterly experiments, has shown that this function compensates so fully for the want of respiration by the lungs, as to enable these animals, in the winter season, to live for an almost unlimited period under the surface of the water."*

"It is scarcely possible" says Professor Daniel, "duly to appreciate, in the vast economy of terrestrial adaptations, the importance of this mechanism, by which gases and vapours rapidly permeate each other's bulks, and become equally diffused. The atmosphere which surrounds the globe consists of a mixture of several aëriform fluids, in certain fixed proportions, upon the proper maintenance of which, by measure and by weight, the welfare of the

* Turner's Elements.

whole organic creation depends. The processes of respiration and of combustion are perpetually tending to the destruction of the vital air, and the substitution of another which is a deadly poison to animal life; and yet, by the simple means which we have described, the poisonous air is not allowed to accumulate, but diffuses itself instantly through surrounding space, while the vital gas rushes, by a counter tendency, to supply the deficiency which the local consumption has created. Hence the invariable uniformity of this mixture, which is one of the most surprising phenomena where all is admirable. The most accurate examination has been made of air which has been taken from localities the most opposed to each other, in all the circumstances which can be conceived to affect its purity; by means of a balloon, from a height of 22,000 feet above the level of the sea; from the surface of the ocean; from the summit of Mont Blanc; from the heart of the most crowded districts of the most populous cities; from within the Polar Circle; and from the Equator; and no difference has been detected in the proportions of its principal constituents."

CHAPTER III.

ON THE IMITATION OF THE NATURAL CONDITIONS OF PLANTS IN CLOSELY GLAZED CASES.

"Homo, Naturæ minister et interpres, tantum facit et intelligit quantum de Naturæ ordine re vel mente conservaverit : nec amplius scit aut potest." — BACON, NOVUM ORGANUM, APHOR. 1.

"The power of man over Nature is limited only by the one condition that it must be exercised in conformity with the laws of Nature."—HERSCHELL.

CHAPTER III.

ON THE IMITATION OF THE NATURAL CONDITIONS OF PLANTS IN CLOSELY GLAZED CASES.

THE science of Botany, in consequence of the perusal of the works of the immortal Linnæus, had been my recreation from my youth up ; and the earliest object of my ambition was to possess an old wall covered with ferns and mosses. To obtain this end, I built up some rock-work in the yard at the back of my house, and placed a perforated pipe at the top, from which water trickled on the plants beneath ; these consisted of *Polypodium vulgare, Lomaria spicant, Lastræa dilatata, L. Filix-mas, Athyrium Filix-fœmina, Asplenium Trichomanes,* and a few other ferns, and several mosses procured from the woods in the neighbourhood of London, together with primroses, wood sorrel, &c. &c. Being, however, surrounded by numerous manufactories and enveloped in their smoke, my plants soon began to decline, and ultimately perished, all my endeavours to keep them alive proving fruitless. When the attempt had been given up in despair, I was led to reflect a little more deeply upon the subject, in consequence of a simple incident which occurred in the summer of 1829. I had buried the chrysalis of

a Sphinx in some moist mould contained in a wide-
mouthed glass bottle, covered with a lid. In watching
the bottle from day to day, I observed that the moisture
which during the heat of the day arose from the mould,
became condensed on the internal surface of the glass,
and returned whence it came; thus keeping the mould
always in the same degree of humidity. About a week
prior to the final change of the insect, a seedling fern and
a grass made their appearance on the surface of the mould.

I could not but be struck with the circumstance of one
of that very tribe of plants, which I had for years fruit-
lessly attempted to cultivate, coming up *sponte suâ* in
such a situation; and asked myself seriously what were
the conditions necessary for its growth? To this the
answer was,—1stly, an atmosphere free from soot; (this I
well knew from previous experience) : — 2ndly, light : —
3rdly, heat :— 4thly, moisture :— and lastly, change of air.
It was quite evident that the plants could obtain light and
heat as well in the bottle as out of it; and that the lid
which retained the moisture likewise excluded the soot.
The only remaining condition to be fulfilled was the change
of air; and how was this to be effected? When I published
my account in the 'Companion to the Botanical Magazine,' I overlooked the law respecting the diffusion of
gaseous bodies, described in the preceding chapter, and
stated that this change was produced by the variations of
temperature causing alternate expansions and contractions
in the air surrounding the plants, and which of course
produced a certain but very limited effect.

Thus, then, all the conditions necessary for the growth
of my little plant were apparently fulfilled, and it remained
only to put it to the test of experiment. I placed the

bottle outside the window of my study — a room facing the north, and to my great delight the plants continued to grow well. They turned out to be *Lastræa Filix-mas* and *Poa annua*. They required no attention, the same circulation of the water continuing; and here they remained for nearly four years, the *Poa* once flowering, and the fern producing three or four fronds annually. At the end of this time they accidentally perished, during my absence from home, in consequence of the rusting of the lid, and the admission of rain water. Long before this occurred, however, I procured for the purposes of experiment some plants of *Hymenophyllum* and *Trichomanes;* and perhaps the most instructive way in which I can communicate the results of my enquiries will be to select a few out of numberless experiments, in the order in which they occurred. To commence with—

1. *Trichomanes speciosum;* (the *T. brevisetum* of most English botanical works). This, the most lovely of our cellular plants, is the most intractable under ordinary methods of treatment. Loddiges, who have had it repeatedly, never could keep it alive; * and Baron Fischer, the superintendant of the botanical establishments of the Emperor of Russia, when he saw the plant growing in one of my cases, took off his hat, made a low bow to it, and said — "You have been my master all the days of my life." Whence then arises the great difficulty of cultivating this plant? It is simply owing to the occasional dryness of the atmosphere. Place the plant in one of my

Killarney fern

* Mr. Mackay, of Dublin, I believe is almost the only person who has succeeded in growing this plant well; and to him I am indebted for my present specimens, and for numerous other kind contributions.

Poa annua — annual meadow grass
Lastraea Filixmas — lacey-type ferns
Hymenophyllum — membranous leaf
Trichomanes — bristle ferns

cases, and thus secure a constantly humid atmosphere around it, and it will grow as well in the most smoky parts of London as on the rocks at Killarney, or in the laurel forests of Teneriffe : —

" Miraturque novas frondes."

This plant lived for about four years in a wide-mouthed bottle, covered with oiled silk, during which time it required no water; but having outgrown its narrow limits it was removed to some rock-work in my largest fern-house, where it now remains, covered with a bell-glass, and occasionally watered.

2. *Hymenophyllum, with one or two species of Junger-mannia and Mosses.* These were planted nine years since, in the bottle in which my first experimental plants sprang up and perished. The soil is a mixture of peat mould, loam and sand, with as much moisture as it would retain when water was poured through it. This same water has served for the nourishment of the plants up to the present time, nor am I at present able to assign any limit to their existence in this state. The mould appears to be as moist and the plants as fresh, as on the day they were enclosed ; and when we reflect upon their independent state, we may, without any great stretch of imagination, carry our minds back to the primæval condition of vegetation, when "the Lord God had not caused it to rain upon the earth, and there was not a man to till the ground. But there went up a mist from the earth, and watered the whole face of the ground."

This will be a fitting place to make mention of a small but most interesting bottle which I received in October, 1837, from Mr. Newman, superintendant of the

Botanic Garden at the Mauritius. The bottle was filled with two or three specimens of a little species of *Gratiola* and of *Cotula,* and tightly covered with painted canvas. The plants were in full flower. I placed them in a window with a southern aspect : they remained in vigour for six or seven weeks, when one after the other declined, and eventually all perished without ripening any seed, in consequence of the too great humidity of the atmosphere. Before this took place I observed, as in my first experiment, several seedling ferns making their appearance between the internal surface of the glass and the mould, and therefore allowed the bottle to remain in the same situation, which it has occupied to the present time, the cover never having been removed; and it is now a truly beautiful object. The upper part of the bottle is completely filled with the fronds of two species of *Adiantum* and one or two other species of ferns, and the lateral surface of the mould is densely coated with seedling ferns in all stages.

We may learn a few useful lessons from this little bottle. We see how abundant the seeds of ferns are, and how easy it would be to procure many species of these plants from distant countries, by collecting here and there a handful of the surface-mould, and, at any convenient season, placing this in a condition favourable for their development. To those cavillers who are continually questioning me as to the utility of ferns in creation, I answer, that one of the useful purposes which they serve, in common with numerous other cellular plants, is that of furnishing mould in situations where other plants of a higher order could not at first grow; and this is effected in a two-fold way — by the decay of their fronds and the action of

their roots. Mr. Webster, in his account of the voyage
of the Chanticleer, states that in the course of his ramble
in the Island of St. Catherine's, when gathering ferns, he
was particularly struck by observing that each plant had
formed for itself a bed of fine mould, several inches in
depth and extent, whilst beyond the circle of its own im-
mediate growth was naked rock; and this appeared so
general that he could not help attributing the extraordinary
circumstance to the disintegrating power of their fibrous
roots,* which penetrated every crevice of the rock, and,
by expanding in growth, appeared to split it into the
smallest fragments.

Having determined the complete success of this mode
upon more than a hundred species of ferns, and my ideas
having a little expanded, I built a small house about eight
feet square, outside one of my stair-case windows, facing
the north; and, proceeding from ferns to those plants
which live in their company, filled it with a mixed vegetation.   This is called—

3. *The Tintern-Abbey House;* from its containing in the
centre a small model, built in pumice and Bath stone, of
the west window of Tintern Abbey.   The sides are built
up with rock-work to the height of about five feet, and a
perforated pipe runs round the top of the house, by means

---

* The *Opuntia*, or Prickly Pear, when placed in fresh fields of lava,
which, in the ordinary course of nature,—i. e. by the successive growth
and decay of lichens, mosses, and other cellular plants, would require
a thousand years to become fertile, renders them capable of being con-
verted into vineyards in the course of thirty or forty; and this by the
comminuting action of its roots.   Indeed, in all cases, the formation of
mould may be traced to the double cause of the decay of dead vegetable matter and the splitting power of living roots.

of which I can rain upon the plants at pleasure. In the middle of summer the sun shines into this house for about one hour only in the morning, and about the same time in the evening, but not at all during the winter. There is no artificial heat. It contains at present about fifty species of British, North American, and other hardy ferns, *Lycopodium denticulatum, lucidulum* and *clavatum*, and the following flowering plants—*Linnæa borealis, Oxalis Acetosella, Primula vulgaris, Digitalis purpurea, Cardamine flexuosa, Lonicera Periclymenum, Meconopsis cambrica, Geranium robertianum* var. *flore albo, Dentaria bulbifera, Paris quadrifolia, Mimulus moschatus, Linaria Cymbalaria, Lamium maculatum,* and several others. All these flower well, but the atmosphere is too moist, and there is too little sun for them to ripen seed; with the exception of the *Mimulus*, the *Oxalis*, and the *Cardamine*, which latter grows with great luxuriance, and furnishes throughout the year a most grateful addition to the food of a tame Canary bird. The *Rhapis flabelliformis* and *Phœnix dactylifera* bore the cold during three winters in this house, when I was obliged to remove them in consequence of their size. A double white Camellia was also planted, about four years back, and blossomed tolerably well for three successive springs, but was killed by the last severe winter. In a cold house like this, but with an eastern or western aspect, so as to admit more solar light, I believe that Camellias would flower beautifully, and be far less likely to suffer from the winter's cold. The influence of light in enabling plants to withstand cold is far too little attended to, and in all cases where it is necessary to protect delicate plants in winter, light should be admitted, if possible. I shall next mention —

4. *The Alpine Case.*   *Azalea procumbens, Andromeda
tetragona, A. hypnoides, Primula minima, P. helvetica,
Soldanella montana, S. alpina, Eriophorum alpinum* and
a few others, were the contents of my first alpine case.
As I thought there would not be sufficient light at any of
my windows, I placed the case on the roof of the house,
and in the following spring all the plants flowered well
except the Andromedas.   Forgetting that an alpine summer is not so long as ours, I allowed the plants to remain
fully exposed to the sun for the whole year, owing to
which they became so exhausted that some died, and but
few flowered in the ensuing spring.   Warned by this, in
my succeeding experiments on this interesting tribe of
plants, I remove the case after their flowering into the
coldest and most shady place I can find, until the following winter, when they resume their old position.   In this
way they flourish much better, but it is impossible to do
them full justice, as we cannot give them the perfect rest
which they require.

5. *Drawing-Room Case.*   This case is at present filled
in the bottom with two or three small Palms, some ferns,
two or three species of *Lycopodium,* and several bulbous
roots.   Within, and along the roof of this case, runs a
perforated bronze bar, from which are suspended small
pots, containing *Mammillaria tenuis,* two or three species
of *Cactus,* and one or two Aloes.   In such a case as this
it is easy to grow bog plants in the bottom and succulents
at the top, these last never receiving any moisture but in
the state of vapour, and that more abundantly when they
most want it, viz. in the heat of summer.   The distance
between the surface of the mould at the bottom and the
suspended plants does not exceed eighteen inches.   The

case stands in the window of a room with a southern aspect, and the thermometer in summer frequently rises to 110°, and even higher. This case requires occasional watering.

6. *Small Bottle with Mammillaria tenuis, a species of Cactus, and two or three fleshy species of Euphorbia.* This stands under the drawing-room case. The plants have been enclosed four years; the mould consisting of very sandy loam. No water has been given since they were planted, and all are in a state of perfect health, although now outgrowing their narrow bounds.

7. *Crocuses and Winter Aconites.* Two cases were filled with roots of these plants; the one placed outside a window with a southern aspect, where there was plenty of light, but no artificial heat; the other in a warm room, where the light was very deficient. The plants in the former case exhibited a perfectly natural appearance,— their flowers were abundant and well coloured; while in the latter the leaves grew very long and pale, and not a single flower was produced.

8. *Crocuses with Artificial Light.* A case fitted up precisely as the two preceding was placed on my staircase, close to a gas lamp. The plants were covered during the day with a thick dark cloth, so as effectually to exclude day-light, and as soon as the gas was lighted the cloth was removed. The plants were thus exposed from five to eight hours daily to the influence of artificial light, accompanied with some degree of heat, while the remainder of the twenty-four hours was spent in a state of rest. The plants grew very well, the leaves not so much drawn up as those in the warm room, and the color more intense One root flowered, the color of the flower being blue.

D

9. *Case with Spring Flowers.* In order to have a gay assemblage of flowers, I filled a case about three feet by one with the following plants, viz., *Primula sinensis, P. nivalis, Scilla sibirica, Cyclamen Coüm, Ornithogalum Sternbergii, Gagea lutea, Ganymedes pulchellus,* and three or four varieties of Crocus, interspersed with little patches of *Lycopodium denticulatum.* This case was placed, about the end of February, outside a window with a southern aspect. It is not, I believe, possible to see these plants to such advantage in any ordinary garden. Here, undisturbed either by wind or rain, their flowers are developed in the greatest luxuriance ; and most of them continue for two or three months,* realising the beautiful description of Catullus :—

> " Ut flos in septis secretus nascitur hortis
> Ignotus pecori, nullo contusus aratro
> Quem mulcent auræ, firmat sol, educat imber
> Multi illum pueri, multæ optavere puellæ."

10. *Fairy Roses.* I prócured two of the smallest varieties of Fairy Rose, planted them in two tubs, in some good loam, with broken pots at the bottom, and then covered them with bell-glasses, the diameter of which was rather smaller than that of the tubs, and placed them outside a window facing the south, where they have now remained three years. These plants are as nearly as possible in their natural condition, very seldom requiring water, as the rain which falls runs over the glass through the

---

* The *Chorizema ilicifolium,* if placed in such a situation in the beginning of May, will continue to flower for four or five months ; and cut flowers will last twice or thrice as long as in ordinary rooms.

mould. They begin to flower early in the spring, and continue for four or five months in great beauty, nothing more being required than to give them an occasional pruning.

It would be waste of time to detail any more of these minor experiments, and I shall therefore conclude by giving a short description of my largest experimental house. My object in this building was to obtain as many varied modifications of the natural conditions of plants as it was possible to procure in the small space to which I was confined.

The greatest length is twenty-four feet, width twelve feet, and extreme height eleven feet :—

" Exiguus spatio, variis sed fertilis herbis."

By building up rock-work to within a foot of the glass, and by varying the surface in every possible way, very different degrees of heat, light and moisture, are apportioned to the various plants. The house is heated in the winter by means of hot-water pipes, which preserve the lower portion during that season at a much higher temperature than the upper ; the latter however has the advantage in the height of summer. The range of the thermometer throughout the year in the lowest part is between 45° and 90°, whilst at the top it is between 30° and 130°. Thus we procure, in a space not exceeding ten feet, an insular, and what may be called an excessive climate. There is no sunshine from the end of October to the end of March. In the lower portion are planted the following Palms :—
Phœnix dactylifera, P. leonensis, Rhapis flabelliformis, R. Sierotsik, a small but beautiful species from Japan, *Chamœrops humilis, Seaforthia nobilis, Cocos botryophora,*

Sabal palmatus, Latania borbonica, and one or two others. Among the ferns we have *Asplenium præmorsum,** remarkably fine, *Diplazium seramporense* † (the *Asplenium pubescens* of Link), *Didymochlæna sinuosa,* and more than a hundred other species. Of Scitamineous plants, of which there are ten or a dozen species, the *Calathea zebrina* is the most conspicuous. The *Caladium esculentum,* and numerous other plants which do not require much sun, likewise grow in this part of the house. In the upper region are numerous species of *Aloë, Cactus, Bilbergia, Begonia,* &c. &c. Two or three varieties of rose likewise flower here, but neither so well nor so freely as in the cases already described. In hot summers the *Mimosa pudica* flowers freely, as do one or two species of *Passiflora.* In the intermediate spaces are *Disandra prostrata, Fuchsias,* and various other plants. From the roof are suspended numerous succulents and Orchideous epiphytes, but the temperature falls too low in the winter, and rarely rises sufficiently high in the summer, for these *splendid things without a foundation,*‡ so that they rarely flower. In a large vessel containing about twenty gallons of water *Papyrus elegans* grows very well, as does *Vallisneria spiralis,* and some other aquatics. In addition to this great variety of living forms, this house contains a large and fine collection of *Lepidodendra, Calamites,* &c.

---

* This is a valuable plant for such a house, as each frond lasts three or four years in perfection.

† This plant, which had been sterile at Loddiges for fifty years, produced a frond two years ago covered with fructification.

‡ The meaning of the name given to them by the South Sea Islanders. —*Williams.*

&c., which, when compared with their recent types, the *Lycopodia* and *Equiseta*, are truly—

———— " of aspect that appears
Beyond the range of vegetative power."

Such are some of the results obtained in a temperate climate, and there cannot, I think, be a doubt that in tropical countries the application of this same plan might be equally striking and beneficial.

In ordinary horticulture a great deal is effected by closely imitating the natural conditions of plants. Thus my friend Dr. Royle, who has paid especial attention to this subject, informed me that there were certain plants in his garden at Saharunpore, which he could only keep alive by surrounding them with small trees and shrubs, so as to give them a moister atmosphere than they could otherwise have obtained; and he mentions in his beautiful work, the 'Illustrations of the Flora and Fauna of the Himalayas,' a striking example of this kind.—" To show the effects of protection and culture *Xanthochymus dulcis* may be adduced as a remarkable instance. This tree, which is found only in the southern parts of India, and which would not live in the more exposed climate of Saharunpore, exists as a large tree in the garden of the King of Delhi; but here, surrounded by the numerous buildings within the lofty palace wall, in the midst of almost a forest of trees, with perpetual irrigation from a branch of the canal which flows through the garden, an artificial climate is produced, which enables a plant even so sensitive of cold as one of the *Guttiferæ* to flourish in the open air of Delhi, where it is highly prized, and reported to have milk thrown over its roots, as well as its fruit protected from plunder by a guard of soldiers."

Supposing ourselves in a hot and dry country, let us see what may be done by surrounding our plants with glass and lowering the temperature, if required, by means of the evaporation of water from the external surface. We shall be enabled in this manner, as with the wand of a magician, to turn a desert into a paradise. The probable results cannot be better described than by copying the beautiful description of the palm-groves given us by Desfontaines, in his ' Flora Atlantica.'

" Palmeta radiis solis impervia, umbram in regione calidissima hospitalem incolis, viatoribus, æque ac animantibus ministrant. Eorum denso sub tegmine absque ordine crescunt aurantia, limones, punicæ, oleæ, amygdali, vites, quæ cursu geniculato sæpe truncos palmarum scandunt. Hæ omnes fructus suavissimos, licet obumbratæ ferunt; ibique mira florum et fructuum varietate, pascuntur oculi; simulque festivis avium cantilenis, quas umbra, aquæ, victus illiciunt, recreantur aures."*

There are many other situations where these cases would be useful, as on ship-board or in other places where there

* " These palm groves, being impervious to the sun's rays, afford a hospitable shade both to man and other animals in a region which would otherwise be intolerable from the intense heat. And under this shelter the orange, the lemon, the pomegranate, the olive, the almond, and the vine, grow in wild luxuriance, producing, notwithstanding they are so shaded, the most delicious fruit. And here, while the eyes are fed with the endless variety of flowers which deck these sylvan scenes, the ears are at the same time ravished with the melodious notes of numerous birds, which are attracted to these groves by the cool springs and the food which they there find."—*Kidd's Bridgewater Treatise.*

exists a necessity for economizing water, or in very cold countries, where it is equally necessary to make the best use of the little sun they possess and to protect the plants from cold winds. The cabbages of Iceland and Labrador would surely exceed their present size of about one or two inches in diameter if thus protected.

To conclude with a few general observations.

The advantages of this method of growing plants consist, first, in the power we possess of freeing or sifting the air from all extraneous matters ; — then of imitating the natural condition of all plants, as far as the climate we are living in will enable us so to do ; and of maintaining this condition free from those disturbing causes to which plants are oftentimes subjected from sudden variations of weather. They are preserved of course from the excess or deficiency of moisture, and, owing to the perfectly quiet atmosphere with which they are surrounded, they are able, like man, to bear extremes of heat and cold with impunity, which in ordinary circumstances would destroy them. The experiments of Sir Charles Blagden and others, in heated ovens, are well known, and the performances of Chaubert are familiar to most of my readers. In these instances the immunity is owing to the aqueous exhalations from the surface of the body remaining undisturbed, and acting as a protecting shield. In like manner the *Trichomanes* lived for three years, in a window with a southern aspect, exposed continually to a heat which, without the glass, would have destroyed it in a single day. With respect to cold, the concurrent testimony of all arctic voyagers proves that no inconvenience is felt, even at 70° below zero, provided the air be perfectly still ; but, that if wind arose, although the thermometer generally rose rapidly with the wind, the cold

then became insupportable. We need not go to the Pole
for illustrations of this fact. Every one has felt the difference between walking with his face or his back to one of
our east winds in March, those winds which are often so
destructive to vegetation in the open air, but have not the
slightest effect on enclosed plants. I need not say anything respecting the change of air, as, from the contents
of the preceding chapter, it must be obvious that as soon
as any gas is generated within the case different to the atmosphere without, diffusion immediately commences, and
no mode of closing the cases which I have adopted can
prevent this from taking place.

As regards the cases in which these plants are grown,
their shape or size may be adapted to the situations in
which they are to be placed. The best cover for the
smaller ones is, I think, the oiled silk of which bathing-
caps are made, or thin sheet India-rubber. The frames of
the larger cases should be well painted, and the laps so
filled with putty as completely to exclude soot.

Do plants require water in these cases ? — is a question
frequently asked. This depends not only upon the nature of
the plants, but upon the season of their growth. Almost all
ferns, if enclosed in small cases where the water cannot
escape, will continue to flourish for years, and I believe that
a century might elapse without any fresh water being required. Cactuses, and most succulent plants, would be
equally independent. In larger houses, where the surfaces
are very varied, the water will drain from the upper parts,
and fresh supplies will occasionally be wanted. If we wish
our plants to grow with greater or less luxuriance; we have
of course, at all times, the power to give or withhold water.
Numerous plants require to be well supplied with water up

to and during the period of inflorescence, and when the flowering is over to be kept nearly dry. This is easily effected by removing the cover, and allowing the moisture to evaporate by exposure of the case for a short time to the sun. It is desirable that there should be an opening in the bottom of the cases for the purpose of draining off the superfluous moisture, and likewise of giving us the opportunity of washing the mould with lime water should slugs make their appearance, which sometimes occurs. With respect to the mould, it is perhaps best to select that in which the plants which are to be the subject of experiment ordinarily grow ; but this is not a matter of so much moment as is generally imagined. It is a very common impression that great knowledge of Botany is required before any successful attempts at the cultivation of plants in closed cases can be made; now, it must be obvious, from all that has been said, that whether the plant be grown in a closed case or in the open air, the natural conditions must be fulfilled to ensure success. Again, many complain that the enclosed plants frequently become mouldy ; this arises either from excess of moisture or deficiency of light, or a combination of both causes producing diminished vital action, or else from the natural decay of plants. It is very interesting to watch the progress of this. The moment a plant begins to decay it is no longer of any use ; and those small parasitical fungi, commonly called moulds, are some of the means employed by nature in removing that which has now become an encumbrance : — " cut it down, why cumbereth it the ground ? "

A few words respecting the importance of reflecting on what we see around us will with propriety close this chapter.

The simple circumstance which set me to work must have
been presented to the eyes of horticulturists thousands of
times, but has passed unheeded in consequence of their
disused closed frames being filled with weeds, instead of cu-
cumbers and melons ; and I am quite ready to confess, that
if some groundsel or chickweed had sprung up in my bottle instead of the fern, it would have made no impression
upon me : and again, after my complete success with the
ferns, had I possessed the inductive mind of a Davy or a
Faraday, I ought, in an hour's quiet reflection, to have anticipated the results of years.   I should have concluded
that all plants would grow as well as the ferns, inasmuch
as I possessed the power of modifying the conditions
suited to the wants of each individual.

CHAPTER IV.

ON THE CONVEYANCE OF PLANTS AND SEEDS ON
SHIP-BOARD.

—————————— " The golden boast
Of Portugal and Western India, there
The ruddier Orange and the paler Lime
Peep through their polished foliage at the storm,
And seem to smile at what they need not fear."
                                        COWPER.

CHAPTER IV.

ON THE CONVEYANCE OF PLANTS AND SEEDS ON SHIP-BOARD.

NUMEROUS have been the methods employed in the conveyance of plants to and from distant countries. It is quite unnecessary, however, to enter into any lengthened account of these attempts, as they resolve themselves into two kinds; — the one where the plants are meant to be kept in a passive condition; and the other where means are employed to keep them growing during the voyage.

The best method of keeping plants in a state of rest is the one generally employed, and, I believe, first recommended by Messrs. Loddiges, viz. — the packing them in successive layers of bog-moss (*Sphagnum*), which answers very well for the majority of deciduous trees and shrubs and other plants, when dispatched at the termination of their active season. For the package of Cactuses and other succulent plants, Messrs. Loddiges recommend the driest sand, all vegetable matters being injurious.

But by far the greater number of plants require to be kept growing during the voyage; and, prior to the introduction of the glazed cases, a large majority of these plants perished from the variations of temperature to which they were subjected, — from being too much or too little watered, — from the spray of the sea, — or, when protected from this spray, from the exclusion of light.   The venerable Menzies informed me that, on his return from his last voyage round the world with Vancouver, he lost the whole of his plants from this latter cause.   Again, if the voyage lasts longer than usual and the water runs short, it is not every one who has the care of plants that will imitate the example of the patriotic M. de Clieux, who, in 1717, took charge of several plants of coffee that were sent to Martinico, and approved himself worthy of the trust.   The voyage being long and the weather unfavourable, they all died but one ; and the whole ship's company being at length reduced to short allowance of water, this zealous patriot divided his own share between himself and the plant committed to his care, and happily succeeded in carrying it safe to Martinico, where it flourished, and was the parent stock whence the neighbouring islands were supplied.

When I reflected upon the above causes of failure, it was obvious that my new method offered a ready means of obviating all these difficulties, so far at least as regarded ferns, and plants growing in similar situations ; and in the beginning of June, 1833, I filled two cases with ferns, grasses, &c., and sent them to Sydney under the care of my zealous friend Capt. Mallard, whose reports on their arrival will be found in the Appendix.

The cases were refilled at Sydney in the month of February, 1834, the thermometer then being between 90° and 100°.   In their passage to England they encountered very varying temperatures.   The thermometer fell to 20° in rounding Cape Horn, and the decks were covered a foot deep with snow. At Rio Janeiro the thermometer rose to 100°, and in crossing the line to 120°. In the month of November, eight months after their departure, they arrived in the British Channel, the thermometer then being as low as 40°. These plants were placed upon the deck during the whole voyage and were not once watered, yet on their arrival at the docks they were in the most healthy and vigorous condition; and I shall not readily forget the delight expressed by Mr. George Loddiges, who accompanied me on board, at the beautiful appearance of the fronds of *Gleichenia microphylla*, a plant never before introduced alive into this country. Several plants of *Callicoma serrata* had sprung up from seed during the voyage, and were in a very healthy state.

My next experiment was with plants of a higher order. Ibrahim Pacha, being desirous of procuring useful and ornamental plants for his garden near Cairo, and at Damascus, I was requested by his agents to select them, and they were sent out in August, 1834, in the Nile steamer, to Alexandria. They arrived quite healthy after a passage of two months.* On a subsequent occasion a case-full of coffee plants was dispatched with the like successful result. It is needless to particularize any more instances, as Messrs. Loddiges† have sent out more than four hundred cases to all parts of the world, with uniform success when the proper conditions were observed; and I believe that the plan, where known, is universally adopted. The French and the English Governments have moreover ordered these cases to be used in their expeditions of discovery; and there are few, I imagine, who will now imitate the ill-timed

* Vide Appendix, D.   † Vide Appendix, G.

economy of Mons. Guillemin, who was sent by the Minister of Agriculture and Commerce at Paris, to Brazil, for the purpose of obtaining information respecting the culture and preparation of the tea-plant, and the introduction of this shrub into France. Mons. G. had personal knowledge of the efficacy of the closed plan, having carried out Camellias to Rio in one of my cases ; and he says that his first plan had been to construct boxes on Mr. Ward's system, but the heavy price* deterred him ; while the safety with which he had brought his fruit-trees† from Europe, in a box with sliding panels, induced him to fix finally on this latter mode of construction.

The results I will give in his own words.—" Very pleasing was the sight to me, when, the day after the Heroine had sailed, (May the 20th, 1839), I beheld my eighteen precious boxes arranged two and two in such a situation as kept them steady and level, permitted them to receive light, and to have the moveable panels closed in bad weather. The vigour of my tea-plants, and the lovely verdure of their foliage, had been generally admired at Rio, and I fondly anticipated the most prosperous results from my expedition. But short-lived was this satisfaction. Two days after heavy north winds drove us off our course, the sea became more boisterous than is usual in these latitudes, and the necessity for closing the ports, lest the spray should irrevocably ruin my plants, caused them a

* The cost of glazing the whole of Mons. G.'s cases would not have exceeded £20.

† Had Mons. G. reflected for one moment upon the different states of the fruit-trees and of the tea-plants,—the former being conveyed at the close, and the latter at the commencement of their active season, — he would not, I think, have acted so unwisely.

great injury by the necessary exclusion of light.   To the latter circumstance I attribute the first deterioration of my plants, especially those more recently set.   When the sea became calmer, and permitted us to open the portholes, the wind sweeping the surface of the waves cast a fine salt-water spray upon my boxes, which doubtless proved highly injurious, since the contents of those chests that were exposed to the wind suffered much more than those of the other side.   By the 11th of June most of the teas had lost their foliage, and the stalks even of several had quite dried up.   Some of the seeds had germinated ; the young shoots were slender, long, blanched, and furnished with a few pale leaves.   By the 2nd of July, in latitude 24° north and longitude 42° west, the strongest shrubs were suffering most severely, while some had sent out suckers, and the young seedlings had assumed a greener tint.   Capt. Cecille took great interest in the safety of my protégés, and, while the leakage of some of the water-casks had compelled him to put the whole ship's crew on a slender allowance of water, he ordered me an increased quantity for the benefit of the tea-shrubs.   The vessel arrived at Brest on the 24th of July, only two months after their departure from Rio, and the shrubs reached Paris in the latter end of August, reduced to 1500 in number, about one-third of the original stock including young seedlings."*   This narrative requires no comment. I believe that not one of the plants would have perished in so short a voyage had they been protected by glass.

Although all persons interested in this matter are pretty well acquainted with the cases in which plants are usually

* I am indebted for this account to Hooker's ' Journal of Botany.'

E

sent on voyages, it may not be amiss to say a word or two
respecting them.   In preparing them for the voyage some
little attention is requisite.   The objects to be attained
are, to admit light freely to all parts of the growing plant,
and to make them sufficiently tight to retain the moisture
within and to exclude the salt water from without.   To
effect the latter purpose the glazed frames should be well
painted and puttied some time before they are required
for use.   The lower part of the case, which contains the
mould, need not be more than 6 or 8 inches in depth ;
and the plants succeed better if planted in the soil, than
in separate small boxes, as in the former case the moisture
is more uniformly diffused.   The soil should be that in
which the plants ordinarily grow, and especial care should
be taken that all superfluous moisture should be drained
off, as luxuriance of growth is not to be desired.   Another
point worthy of great attention is to associate plants of
equal or nearly equal rapidity of growth.   Thus Palms
and coniferous plants will travel well together.   In a case
which arrived at Loddiges, three or four years ago, there
were twenty-eight plants of *Araucaria excelsa* without a
single dead or yellow leaf upon them.   If, in this case,
some free-growing plant had been introduced, the proba-
bility is that all the Pines would have perished, in consequence of the rampant plant occupying all the internal
surface of the glass, and excluding the light from the
others.   A great number of plants will travel well in these
cases, if merely suspended from the roof, — such as numerous species of *Orchideæ*, Cactuses, and other succulent
plants.

When on board, all the care which is requisite is to keep
the plants constantly in the light, to remove incrustations

of salt or dirt, and immediately to repair any damage done to the glass, either with fresh glass, if on board, or with tin or wood.

Although I have stated, and truly, that plants in these cases will bear great variations of temperature with impunity, it does not follow that all plants will bear long continued severe cold.   Care should therefore be taken that all tropical plants should be despatched so as to arrive in this country in warm weather.   It has not unfrequently happened that cases full of precious plants, which have reached the Land's End (Cornwall) in a vigorous condition, after a voyage of several months, have perished from the length of time occupied in beating up Channel in the depth of winter.

With respect to the conveyance of seeds. All those which from their oily nature, peculiarity of constitution, or from any other cause, do not long retain their vegetative properties, are best sown in the mould as soon as they are ripe, and will travel in this way with perfect safety, either among other plants or in cases by themselves.   Thus a great number of plants of *Seaforthia nobilis* were introduced into England, by seeds sown in the cases in New Holland; and I am certain that all the fine timber trees and *Coniferæ* of the Himalaya mountains might thus easily be imported into this country.

As to other seeds, the plan which is now found to be the most successful having been published more than seventy years ago by the celebrated John Ellis, I cannot do better than detail it in the words of the author, and I am induced to do so for two reasons, — to render my subject more complete, and to do justice to the memory of a great man,

whose clear account has been so strangely overlooked by
modern writers.—

" Our seedsmen are much distressed for a proper method
to keep their seeds sound and in a state of vegetation,
through long voyages.   Complaints are made, that when
the seeds arrive in the East Indies, and often in the West
Indies, few of them grow, but that most of them are full
of insects, or what they term weevilly.

" This seems to proceed from the damp and putrid heat
of the hold, or too long confinement in close warm air,
which brings these animals to life, which soon begin to
prey on the inside of these seeds ; and those seeds which
are oily turn rancid.   The putrid penetrating steam that
strikes every one upon opening the hatches of a full
loaded ship's hold, after a long voyage, — it is this that
does the mischief to seeds.   This vapour, as the excellent
Dr. Hales observes, will soon become fatal to vegetable
substances, as well as animals.

" When the cavalry of our army in Germany was under
the necessity of being supplied with hay from England,
the difference was but too manifest between the hay which
had been but a month on board and fresh hay that had
never been confined in the hold of a ship.

" Experiments have been made on the best hemp from
Russia, and hemp of English growth, by persons belong-
ing to the navy of great credit and honour, and the difference in the strength is amazing ; the length of the voyage
from Russia, with the very close package that is necessary
to stow that article on board, raises such a heat as to show
evident signs of putrefaction begun, which must weaken
the strongest vegetable fibre.

" To illustrate this further, in an instance of the different manner of packing and stowing seeds for a long voyage, which has lately come to my knowledge, and may be of use, as it not only points out the error, but, in some measure, how to avoid it.

"A gentleman going to Bencoulen, in the island of Sumatra, had a mind to furnish himself with an assortment of seeds for a kitchen-garden; these were accordingly packed up in boxes and casks, and stowed with other goods in the hold of the ship. When he arrived at Bencoulen he sowed his seeds, but soon found, to his great mortification, that they were all spoiled, for none of them came up. Convinced that it must be owing to the heat of the ship's hold, and their long confinement in putrid air, and having occasion to return to England, he determined in his next voyage thither to pack them up in such a manner, and to place them so, as to give them as much air as he could, without the danger of exposing them to salt-water; and therefore put the smaller seeds into separate papers, and placed them among some clean straw in a small close net, and hung it up in his cabin; and the larger ones he put into boxes, stowing them where the free air could come at them and blow through them; the effect was, that, as soon as he arrived at Bencoulen, he sowed them, and in a little time found, to his great satisfaction, that they all grew extremely well. It is well known to our seedsmen that, even here at home, seeds kept in close warehouses and laid up in heaps frequently spoil, unless they are often sifted and exposed to the air. Seeds saved in moist cold summers, as their juices are too watery, and the substance of their kernels not sufficiently hardened to due ripeness, are by no means fit for exportation to warmer climates.

" Our acorns, unless ripened by a warm summer, will not keep long in England : those acorns which are brought from America, and arrive early in the year, generally come in good order, owing to their juices being better concocted by the heat of their summer; and are not apt to shrivel, when exposed to the sun, as ours are.

" These hints are given to show how necessary it is to take care that the seeds we send should be perfectly ripe and dry."*

* ' Directions for Captains of Ships, Sea-Surgeons, and other curious persons who collect Seeds and Plants in distant countries, in what manner to preserve them fit for vegetation.'—*John Ellis, London*, 1770.

CHAPTER V.

ON THE APPLICATION OF THE CLOSED PLAN IN IMPROVING
THE CONDITION OF THE POOR.

E'en in the stifling bosom of the town
A garden, in which nothing thrives, has charms
That soothes the rich possessor ; much consoled,
That here and there some sprigs of mournful mint,
Of nightshade, or valerian, grace the well
He cultivates.   These serve him with a hint
That Nature lives ; that sight-refreshing green
Is still the livery she delights to wear,
Tho' sickly samples of the exuberant whole.
What are the casements lined with creeping herbs,
The prouder sashes fronted with a range
Of orange, myrtle, or the fragrant weed,
The Frenchman's darling ? are they not all proofs
That man, immured in cities, still retains
His inborn, inextinguishable thirst
Of rural scenes, compensating his loss
By supplemental shifts, the best he may ?
The most unfurnished with the means of life,
And they that never pass their brick-wall bounds
To range the fields, and treat their lungs with air,
Yet feel the burning instinct : over head
Suspend their crazy boxes, planted thick,
And water'd duly.   There the pitcher stands
A fragment, and the spoutless tea-pot there ;
Sad witnesses how close-pent man regrets
The country ; with what ardour he contrives
A peep at Nature, when he can no more.

THE TASK, BOOK 4.

CHAPTER V.

ON THE APPLICATION OF THE CLOSED PLAN IN IMPROVING THE CONDITION OF THE POOR.

AMONG the numerous useful applications of the glazed cases, there is one which I believe to be of paramount importance, and well deserving the attention of every philanthropist : I mean its application to the relief of the physical and moral wants of densely crowded populations in large cities. Among the members of this population there are numbers, who, either from early associations, or from that love of Nature which exists to a greater or less degree in the bosom of all, are passionately fond of flowers, and endeavour to gratify their taste at no small toil. But before I proceed to show how this may be effected, there is one point of so much importance to the well-being of every individual, that I cannot pass it by without notice. This is the free admission of light * into their dwellings. I have already alluded to the effects of light upon vegetation ; and its influence upon the animal economy, although not so immediately obvious, is not the less striking. This has been proved by the experiments of Dr. Edwards.

---

* " Let in the sun and you shut out the doctor," says an old Italian proverb.

" He has shown that if tadpoles be nourished with
proper food, and are exposed to the constantly renewed
action of water (so that their branchial respiration may be
maintained), but are entirely deprived of light, their growth
continues, but their metamorphosis into the condition of
air-breathing animals is arrested, and they remain in the
form of large tadpoles.   Dr. Edwards also observes, that
persons who live in caves and cellars, or in very dark and
narrow streets, are apt to produce deformed children ; and
that men who work in mines are liable to disease and deformity beyond what the simple closeness of the atmosphere would be likely to produce.   It has recently been
stated, on the authority of Sir A. Wylie, that the cases of
disease in the dark side of an extensive barrack at St. Pe-
tersburgh, have been uniformly, for many years, in the
proportion of three to one to those on the side exposed to
strong light.   On the contrary, the more the body is exposed to the influence of light the more freedom do we
find, *cæteris paribus*, from irregular action and conformation.   Humboldt has remarked that among several nations
of South America, who wear very little clothing, he never
saw a single individual with a natural deformity ; and
Linnæus, in his account of his tour through Lapland, enu-
merates constant exposure to solar light as one of the causes
which render a summer's journey through high northern
latitudes, so peculiarly healthful and invigorating."*

I will now endeavour to show how the glazed cases may
be made subservient to the benefit of the poor, and to
point out how cheaply and easily this may be effected.
A box lined with zinc, and having three or four open-

---

* Carpenter's Physiology.

ings in the bottom, will be required for the reception of
the plants; and glazed frames can be procured anywhere,
well painted and puttied, at about one shilling the square
foot. The plants to furnish it can be procured abundantly in the woods in the neighbourhood of London.
Of these I will mention a few. The common Ivy grows
most beautifully, and can be trained over any part of
the case, agreeably to the pleasure of the owner. The
Primroses, * in early spring, will abundantly repay the
labour of fetching them, continuing for seven or eight
weeks in succession to flower as sweetly as in their native
woods. So likewise does the Wood-sorrel, the Anemone,
the Honeysuckle, and a host of other plants, independently
of numerous species of mosses and of ferns. Some of
these latter are more valuable than others, in consequence
of the longer duration of their fronds, such as *Lastræa
dilatata* and its numerous varieties. There are likewise
many cultivated plants procurable at little or no cost,
which grow without the slightest trouble, such as the *Lycopodium denticulatum*, the common Musk-plant, Myrtles,
Jasmines, &c. All the vacant spaces in the case may be
employed in raising small salads, radishes, &c.; and I
think that a man would be a bad manager who could not,

---

* There is, perhaps, no plant which offers so striking an illustration of
the protection afforded by the glass as the common Primrose. Place,
side by side, in a tub outside any smoky window with an eastern aspect,
and where there is no artificial heat, two roots of Primroses, supplying
them with water if needed. Cover one of these with a glass; and the
difference in flowering is so great, that I cannot illustrate it better than
by comparing it to the difference which takes place in the burning of
charcoal, or any other combustible substance, in oxygen gas and in the
open air.

in the course of a twelvemonth, pay for his case out of its proceeds. These remarks apply chiefly to situations where there is but little solar light. Where there is more sun, a greater number and variety of flowering plants will be found to thrive, such as several kinds of Roses, Passion-flowers, Geraniums, &c., with numerous beautiful annuals, viz.— *Ipomœa coccinea*, the species of *Nemophila*, *Convolvulus*, and a host of others : the vegetation in fact can be diversified in an endless degree, not only in proportion to the differing degrees of light and heat, but likewise by varying the quantity of moisture ; thus, with precisely the same aspect, ferns and bog plants might be grown in one case, and Aloes, Cactuses, Mesembryanthemums, and other succulent plants in another.

These cases form the most beautiful blinds that can be imagined, as there is not a window in London which cannot command throughout the year the most luxuriant verdure. The condensation of the moisture upon the colder surface of the glass effectually obscures the view from without, and at the same time admits far more light than is allowed to enter by ordinary blinds. Nothing can be conceived more cheerful than the appearance of rooms thus furnished. As these cases become more general among the higher and middle classes, a new field of healthful industry will thus be opened to the poor, who might not only be employed in procuring plants for these cases from the country, but whose ingenuity might be called into play in executing various models of old towers, ruins, &c., in sand-stone, chalk, or other suitable material, which, at the same time that it served to ornament the case, would afford a suitable place for the growth of little Sedums and any plants that require less moisture than those which are planted in the mould. I need how-ever say no more upon this point, as these various appli-cations must be sufficiently obvious to every one.

But I must here caution the poor against indulging a taste for what are called fancy flowers,—things which this year are rewarded with gold medals, and the next are thrown upon the dunghill. Believing that all human pur-suits ought to be estimated in exact proportion as they tend to promote the glory of God, or the good of man, let us for a moment compare the empty chase after fancy flowers with the legitimate pursuits of horticulture and floriculture. So far from the love of God, and the good of his fellow creatures, being the end and aim of the fancy florist, he values everything in proportion as it is removed from Nature, and unattainable by the rest of mankind. " A long time must elapse ere the world can hope to see a perfect Pansy " ! ! ! says one of these fancy writers. How the world is to benefit by this Phœnix when it does arrive he will of course inform us in his next publication.* Let me entreat the poor to remember that their single talent should be well employed ; † let them learn to estimate things according to their true value, and devote their time and attention to the legitimate pursuits of horticulture and floriculture. It would appear that innumerable plants have been created with latent powers of usefulness for the purpose of exercising the mind, and rewarding the indus-

* That I have not expressed myself too strongly upon this point any one may convince himself, by consulting the history of *Tulipomania.*

† " His virtues walked their narrow round,
    Nor made a pause, nor left a void ;
    And sure th' eternal Master found
    The single talent well employed."—*Johnson.*

try of man, who, by acting in conformity with the laws of
Nature, is enabled to produce the most beneficial results.
Thus, if increased succulence be the point aimed at, the
plants must be the more abundantly supplied with water ;
if increase of flavour, then less water but a larger proportion of sun and light, which latter are to be withheld if
the natural flavour of the plant be too strong. Who could
have imagined from the appearance of the wild carrot or
parsnip, the crab, the celery, or the endive, that all these
would form such important additions to our tables. There
is, in fact, scarcely a vegetable or fruit that owes not a
portion of its excellence to horticultural exertions, directed by Science ; and there cannot be a doubt that many
now-neglected weeds will hereafter become valuable subjects for the horticulturist. And so, with respect to floriculture, that man would be fastidious indeed who would
not appreciate and enjoy the increased beauty and fragrance
of a double Rose or fine Stock. I have said quite enough
as to the physical results of these pursuits, and will endeavour to point out the probable moral effects. The
highest and best feelings of our nature are excited by the
contemplation of the works of God. The Divine Word
has commanded us to " consider the lilies of the field, how
they grow," and the reason assigned is " if God so clothe
the grass of the field which to-day is, and to-morrow is
cast into the oven, how much more shall he clothe you,
oh ! ye of little faith." There is possibly no study which
leads the mind of the pursuer more directly to " the Author
and Giver of all good things," and fills the heart of man
with joy and thankfulness, than the study of that branch
of Natural History which comprehends the vegetable
kingdom. " The infinite variety of forms of the different

species, the nice adaptations of these to their several
functions, the beauty and elegance of a large number, the
singularity of others, but above all their pre-eminent utility
to mankind in every state and stage of life, render them objects of the deepest interest both to rich and poor, high and
low, wise and unlearned; so that arguments in proof of the
Power, Wisdom, and Goodness of God, drawn from the
vegetable kingdom, are likely to meet with more attention,
to be more generally comprehended, to make a deeper and
more lasting impression upon the mind, to direct the heart
more fervently and devotedly to the Maker and Giver of
these interesting beings, than those which are drawn from
more abstruse sources, though really more elevated and
sublime."*   I cannot enforce what I have said better than
by giving a passage from the life of a celebrated traveller.
" Whichever way I turned nothing appeared but danger
and difficulty.   I saw myself in the midst of a vast wil-
derness, in the depth of the rainy season, naked and alone,
surrounded by savage animals, and men still more savage.
I was five hundred miles from the nearest European settle-
ment.   All these circumstances crowded at once upon my
recollection, and I confess that my spirits began to fail
me.   I considered my fate as certain, and that I had no
alternative but to lie down and perish.   The influence of
religion, however, aided and supported me.   I reflected
that no human prudence or foresight could possibly have
averted my present sufferings.   I was indeed a stranger in
a strange land, yet I was still under the protecting eye of
that Providence who has condescended to call himself the
stranger's friend.   At this moment, painful as my reflec-

* Kirby's ' Bridgewater Treatise.'

tions were, the extraordinary beauty of a small moss in fructification irresistibly caught my eye.   I mention this to show from what trifling circumstances the mind will sometimes derive consolation; for though the whole plant was no larger than the top of one my fingers, I could not contemplate the delicate conformation of its roots, leaves and capsules, without admiration.   Can that Being (thought I) who planted, watered, and brought to perfection, in this obscure part of the world, a thing which appears of small importance, look with unconcern upon the situation and sufferings of creatures formed after his own image ?—surely not !   Reflections like these would not allow me to despair.   I started up, and disregarding both hunger and fatigue, travelled forwards, assured that relief was at hand, and I was not disappointed." — *Park's Travels in Africa.*

---

* " The moss which engaged Mungo Park's attention so much in the desert is the *Fissidens bryoides,* as I have ascertained by means of original specimens given to me by his brother-in-law, Mr. Dickson."—*Sir W. J. Hooker.*

CHAPTER VI.

ON THE PROBABLE FUTURE APPLICATIONS OF THE PRECEDING FACTS.

F

"It will be enough if, after having led the way on a new territory of investigation, we shall select one or two out of the goodly number of instances, as specimens of the richness and fertility of the land."—CHALMER'S BRIDGEWATER TREATISE, 2—74.

CHAPTER VI.

ON THE PROBABLE FUTURE APPLICATIONS OF THE PRECEDING FACTS.

THE application of the closed cases to the illustration of physiological and pathological Botany must be sufficiently obvious to all who are interested in such enquiries. In most of the experiments which have hitherto been undertaken by vegetable physiologists, the results have been rendered liable to some doubt, in consequence of the fancied necessity for the open exposure of plants to air; whereas now the utmost certainty can be obtained. I will content myself with specifying a few of the more important instances, in which the close method of growing plants will be found of practical utility.

1. Observations strictly comparative can now be made on the effects of different soils, manures, &c., in cases divided into several compartments, each compartment being filled with different soils, but with the same plants.

2. To determine the powers possessed by plants of absorbing and selecting various substances by their roots.

3. To ascertain the existence and nature of the deleterious excretions from the roots ; — the poisonous character of these excretions, if they exist, being rendered very

problematical by the circumstance of plants in a state of
nature occupying the same situations for ages.

4. To prove the effects of poisons upon plants.

5. To test the influence of light in protecting plants
from the effects of low temperatures.   This can easily be
proved by filling a case with specimens of one plant, and
by darkening portions of the glass.

In the severe winter which occurred three or four years
ago, the noble plant of *Araucaria excelsa* in the Pinetum
at Dropmore was killed.   I believe that the plant would
not have suffered had light been admitted through the
covering which protected it from the cold; and this could
easily have been effected by means of melon-lights, &c.

6. To determine various important points respecting
those numerous and highly interesting tribes of plants,
which, from their extreme minuteness or fugacious nature,
have hitherto almost eluded observation, but which the
botanist in his study will now be enabled to watch, micro-
scopically if required, during the whole period of their
growth.   To give an instance : — I had been struck with
the published accounts of the extraordinary growth of
Phallus fœtidus, which was said to attain a height of four
or five inches in as many hours.   I procured three or four
specimens in an undeveloped state and placed them in a
small glazed case.   All but one grew during my temporary
absence from home.   I was determined not to lose sight
of the last specimen ; and observing one evening that
there was a small rent in the volva, indicating the ap-
proaching development of the plant, I watched it all night,
and at eight in the morning the summit of the pileus began to push through the jelly-like matter with which it
was surrounded.   In the course of twenty-five minutes it

shot up three inches, and attained its full elevation of four inches in one hour and a half. The entire life of the *Phallus* was four days. Extraordinary as this may appear I believe this rapidity of development to be surpassed by other fungi, as I was informed by Lady Arden, who has paid great attention to the species of this family, of which she has made numerous exquisite drawings, that the lives of some were so brief as scarcely to allow of sufficient time to finish her representations. Marvellous are the accounts of the rapid growth of cells in the fungi; but, in the above instance, it cannot for a moment be imagined that there was any increase in the number of cells, but merely an elongation of the erectile tissue of the plant. Great confusion likewise exists in the determination of the genera and species of this family. Out of one *species*, according to Fries,* no less than eight *genera* have been formed, in consequence of this species being seen in different situations and states of growth.

Lastly, I believe that by means of these cases the scientific naturalist will be assisted in exploring that debatable ground on the confines of the animal and vegetable kingdoms, where in our present state of ignorance, it is often impossible to determine the point at which the one ends or the other begins.

I shall conclude my little work with pointing out the application of the principle to animals and to man; an application far outweighing in importance all that has

---

* This author asserts that out of mere degenerations or imperfect states of *Thelephora sulphurea*, the following genera, all of which he has identified by means of unquestionable evidence, have been constructed, viz. — *Athelia* of Persoon, *Ozonium* of Persoon, *Himantia* of Persoon, *Sporotrichum* of Kunze, *Alytosporium* of Link, *Xylostroma*, *Racodium* of Persoon, *Ceratonema* of Persoon, and some others.—*Lindley's Introd.*

hitherto been done. In my report to the meeting of the British Association at Liverpool, in 1838, I directed the attention of the members to this subject, at the same time expressing my conviction that a great number of animals would live and thrive under the same plan of treatment which had proved so successful with plants. A little reflection will convince us that this idea is not so visionary as it might appear at first sight. It has now been proved by numerous and long-continued experiments, that the air of London, if duly sifted, is perfectly fitted for the respiration of all plants, even of those with the most delicate leaves, such as the *Trichomanes speciosum*, which may in fact be considered a test plant as regards the purity of the air. Now this same condition of the atmosphere, so essential to the well-being and even the existence of such plants, we have it in our power to obtain in large towns; and by warming and moistening the air we can in fact closely imitate any climate upon the face of the earth. It cannot be denied that in a pure and properly regulated atmosphere we possess a remedial means of the highest order for many of the ills that flesh is heir to; and every medical man knows well, by painful experience, how numerous are the diseases which, setting at nought his skill and his remedies, would yield at once to the renovating influence of pure air. The difficulty to be overcome would be the removal or neutralization of the carbonic acid given out by animals; but this in the present state of science could easily be effected, either by ventilators or by the growth of plants in connexion with the air of the room, so that the animal and vegetable respirations might counterbalance each other. The volume of the air with the quantity of vegetable matter required, as compared with the size and rank in creation of the animal, would be a problem well worthy of solution. Experiments of this kind upon any scale might be instituted in the Zoological Gardens, where the moping owl and ivy-mantled tower might be associated. In one of my own houses, about ten feet square, which was filled with small Palms, and sufficiently close for the growth of the most delicate ferns, a robin lived for several months, at the end of which time he escaped in consequence of the accidental opening of the door.

Among the diseases incident to man, which would be most materially benefited by pure air, I shall allude only to two, viz.—measles and consumption. This is not the place to enter into any long discussion on medical points ; but, believing firmly as I do that a properly regulated atmosphere is of more importance in these diseases than all other remedial means, it would have been unpardonable in a work like the present to have passed them over without notice. In the crowded districts of large towns the direct mortality arising from measles is always great, but nothing I believe compared with the numbers that die at various and distant intervals in consequence of neglect during the disease. Nearly all this distress and mortality might be averted were there proper rooms provided for the reception of the children of the poor when labouring under this complaint, or even of communicating it in favorable seasons. With respect to consumption, could we have such a place of refuge as I believe one of these closed houses would prove to be, we should then be no longer under the painful necessity of sending a beloved relative to a distant land for the remote chance of recovery, or too probably to realize the painful description of Blackwood : — " Far away from home, with strangers

around him,—a language he does not understand,—doctors
in whom he has no confidence,—scenery he is too ill to
admire,—religious comforters in whom he has no faith,—
with a deep and every day more vivid recollection of
domestic scenes,— heart-broken, — home-sick, — friendless
and uncared for,—he dies."

In concluding my little work I feel that apologies are
due for its many imperfections.   The unremitting toil of
general medical practice allows of little time for scientific
enquiries ; and I shall be satisfied if the facts which have
been stated, and the hints which have been thrown out,
excite the attention of those who possess more knowledge,
time and means, than myself. Deeply convinced of the great
practical utility and high importance of these researches,
I hope yet to see the day when, in our universities and
great public schools, the study of Natural History will be
deemed at least as worthy of attention as an ode of Pindar,
or a proposition of Euclid; and the students no longer
presented—

> " with an universal blank
> Of Nature's works, to them expunged and rased,
> And wisdom at one entrance quite shut out."

APPENDIX.

APPENDIX.

(A)

Copy of a Letter to DAVID DON, *Esq., read before the Linnean Society of London, June 4th,* 1833.

Wellclose Square,
June 4, 1833.

My Dear Sir,

The difficulty of conveying ferns from foreign countries has long been matter of regret to the cultivators of that most interesting family of plants.   About three years ago I was led to make some experiments upon the subject, in consequence of noticing a seedling of *Aspidium Filix-mas,* and one of *Poa annua,* on the surface of some moist mould in a large bottle, in which I had buried the chrysalis of a *Sphinx.*   Curious to observe how vegetation would proceed in so confined a situation, I placed the bottle, loosely covered with a tin lid, outside one of my windows, with a northern aspect.   This cover allowed a sufficient change of air for the preservation and development of the plants, and, at the same time, prevented the evaporation of the moisture within. In the bottle these plants remained for more than three years, during which time not one drop of water was given to them, nor was the cover removed. The *Poa* flowered the second year, but did not ripen its seeds; and about five or six fronds of the *Aspidium* were annually developed, but neither thecæ nor sporules were produced. These plants accidentally perished, from the rusting of the lid and the consequent admission of rain, which caused them to rot. During the last twelvemonth I have tried this method with more than thirty species of ferns, with uniform success. Many other plants which grow in moist situations will succeed equally well when treated in this way. To mention one instance: I transplanted some roots of *Listera Nidus-avis* about three weeks ago. Those which I placed in my fern-boxes grew most rapidly, while the remainder, treated in the usual manner, completely withered away. I have the pleasure of submitting two of my boxes to the inspection of the Linnean Society. My valued friend Capt. Mallard, whose active zeal in the cause of Science is well known to many Fellows of the Linnean as well as of the Zoological Society, has engaged to convey these boxes on an experimental voyage to New Holland; and I hope, on his return, to find that they have not lost their character by being transported.

I am, My Dear Sir,

Yours very truly,

N. B. WARD.

To David Don, Esq.

(B)

Copy of a Letter from CHARLES MALLARD, *Esq., R.N. to the Author.*

Hobart Town,
November 23, 1833.

Sir,

You will, I am sure, be much pleased to hear that your experiment for the preservation of plants alive, without the necessity of water or open exposure to the air, has fully succeeded.

The two boxes entrusted to my care, containing ferns, mosses, grasses, &c., are now on the poop of the ship, (where they have been all the voyage); and the plants, (with the exception of two or three ferns which appear to have faded), are all alive and vigorous.

During the very hot weather near the equator, I gave them once a light sprinkling of water, and that is all they have received during the passage.

All the plants have grown a great deal, particularly the grasses, which have been attempting to push the top of the box off.

I shall carry them forward to Sydney, according to your instructions, and have no doubt of delivering them into the hands of Mr. Cunningham in the same flourishing state in which they are at present.

Allow me, in conclusion, to offer to you my warm congratulations upon the success of this simple but beautiful discovery for the preservation of plants in the living state upon the longest voyages ; and I feel not a little pride in having been the instrument by which the truth of your new principle has been fully proved by experiment.

I am, Sir, &c. &c.

CHARLES MALLARD.

To N. B. Ward, Esq.

————

(C)

Copy of a Letter from CHARLES MALLARD, Esq., R.N., *to the Author.*

Sydney,
January 18, 1834.

Sir,

I have the happiness to inform you that the plants contained in the two glazed cases entrusted to my care, were landed here at the Botanical Garden about three weeks ago, nearly the whole of them alive and flourishing. They have since been transplanted by Mr. McLean, who has charge of the garden in the absence of Mr. Cunningham (gone to New Zealand botanizing), and are all doing well.

The complete success of your interesting experiment has been decidedly proved; and whilst offering you my congratulations upon this agreeable result, I cannot but feel some little degree of pride and pleasure in having been the instrument selected to put to the proof so important a discovery to the botanical world.

I am, Sir, &c. &c.

CHARLES MALLARD.

To N. B. Ward, Esq.

————

(D)

Copy of a Letter from MR. TRAILL *to the Author.*

Cairo,
April 30, 1835.

Sir,

I beg to acknowledge the receipt of your letter of
the 2nd ult., wherein you request information as to the state
of the plants sent by you in the Nile steamer.*   The collection consisted, I believe, of 173 species, contained in
six glazed cases, two of which only were forwarded to me
from Alexandria.   The one which you mention as having
been fitted up with talc, together with three others, were
sent on to Syria † immediately on their arrival in Alexandria, so that I had no opportunity of seeing them.   I have,
however, the pleasure to inform you that the Egyptian
portion of the collection was received here in the very best
condition : the plants, when removed from the cases, did
not appear to have suffered in the slightest degree ; they
were in a perfectly fresh and vigorous state, and, in fact,
hardly a leaf had been lost during their passage.   Your
plan, I think decidedly a good one, and ought to be made
generally known.

I am, Sir, &c. &c.

J. TRAILL.

To N. B. Ward, Esq.

---

* In August, 1834.

† These cases were seen by Col. Higgins of the Engineers, in the
garden of the Seraglio, at Beyrout, at the late evacuation of that place
by the Egyptians.

List of Plants contained in the two cases sent to Egypt.

Achras Sapota

Adenoropium panduræfolium

Aleurites moluccana

Alpinia nutans

Anona Cherimolia

Arenga saccharifera

Bignonia venusta

Bombax Gossypium

Brexia spinosa

Calathea zebrina

Caryota urens

Cedrela odorata

Cinnamomum aromaticum

Cinnamomum zeylanicum

Combretum comosum

Croton variegatum

Curcuma longa

Cycas revoluta

Dalbergia scandens

Diospyros cordifolia

Diospyros edulis

Diospyros Embryopteris

Doryanthes excelsa

Dracæna edulis

Dracæna ferrea

Erythrina crista-galli

Eugenia Pimenta

Euphoria Litchi

Ficus elastica

Flacourtia cataphracta

Franciscea uniflora

Jonesia pinnata

Ixora coccinea

Latania borbonica

Maranta arundinacea

Maranta bicolor

Melastoma Fothergilla

Menispermum Cocculus

Melaleuca Cajuputi

Mimusops Elengi

Morus tinctoria

Oreodoxia regia

Pandanus odoratissimus

Passiflora racemosa

Piper Betle

Piper nigrum

Psidium chinense

Terminalia angustifolia

Uvaria odoratissima

Vanilla planifolia

Zingiber officinale

(E)

Extrait du Rapport de la Commission chargée, sur l'invitation de M. le Ministre de la Marine, de rédiger des instructions pour les observations scientifiques à faire pendant le voyage des corvettes de l'Etat l'Astrolabe et la Zélée, sous le commandement de M. le CAPITAINE DUMONT d'URVILLE.

(*Commissaires,* MM. de MIRBEL, CORDIER, de BLAINVILLE, de FREY-CINET, SAVARY.)

Comptes Rendus de l'Academie des Sciences,
7 Aout, 1837.

" Un autre appareil, inventé pour le transport des plantes par M. Nath. Ward, de Londres, offre encore plus de chances de succès que celui de Luschnath ; mais il ne remplit sa destination qu'à la condition que, pendant sa traversée, il restera exposé à l'action de la lumière et n'éprouvera aucune avarie grave. Cet appareil, que nous appellerons *serre de voyage,* consiste en une caisse allongée, surmontée d'un toit vitré, formé par deux châssis ajustés de manière à faire un angle aigu. Les deux petits côtés de la caisse dépassent sa base de deux à trois centimètres, servent de support à tout l'appareil ; et, s'élevant en angle aigu au-dessus de l'ouverture de la caisse, ferment les deux côtés du toit. L'un des châssis est à poste fixe ; l'autre, retenu par quelques vis, se place ou s'enlève à volonté, mais il doit fermer exactement la boîte tant que dure le voyage : alors la parfaite clôture de toutes les parties est de rigueur. Des traverses en bois, de quatre à cinq cent-imètres de large, à la distance l'une de l'autre de sept à huit centimètres, s'ajustent avec la partie inférieure et

G

supérieure de chaque châssis, et servent à la fois à lui
donner de la solidité et à soutenir les verres, qui sont
petits, très épais, à recouvrement comme les tuiles d'un
toit, et mastiqués dans toutes leurs jointures.

" La grandeur des *serres de voyage* peut varier, mais
pour qu'elles ne gênent point les matelots dans l'exécution
des manœuvres, ce qui finalement compromettrait l'existence des plantes, on a soin de les réduire à de petites dimensions. On y trouve d'ailleurs cet autre avantage, qu'il
est plus facile de les rendre imperméables à l'air et à l'eau.
Généralement parlant, les plus grandes dimensions qu'il
convient de leur donner, sont les suivantes, et peut-être
vaut-il mieux rester un peu au-dessous de ces mesures que
les dépasser.

9 décimètres de longueur.
7   „   de hauteur.
5   „   de largeur.

" La profondeur de la caisse, abstraction faite du toit, ne
peut guère être moindre de deux décimètres six centimètres, quelles que soient les dimensions de l'appareil.

" Il est bien entendu que les planches qu'on emploiera,
seront d'un bois solide et sec, et qu'on les ajustera avec le
plus grand soin. On les recouvrira en dehors de plusieurs
couches de couleur à l'huile. Des poignées de fer qui
tourneront dans des pitons, seront attachées solidement
aux deux petits côtés, à la hauteur de trois décimètres en-
viron. Les deux châssis vitrés seront mis chacun à l'abri
des accidents, sous un fort grillage à petites mailles, sou-
tenu par plusieurs tringles de fer assez épaisses pour résis-
ter à des chocs d'une certaine rudesse.

" Quand on veut garnir la *serre de voyage,* on enlève le
châssis mobile, on met au fond de la caisse une épaisseur

de trois à quatre centimètres de terre argileuse, laquelle a
été d'abord humectée, malaxée, battue, et ne contient plus
d'eau sensiblement *mouillante;* et l'on couvre cette couche
d'une terre de bonne qualité, ni trop forte ni trop légère et
bien ameublie. Les végétaux sont placés dans ce sol,
tantôt à racine nue, tantôt à racine en motte revêtue de
mousse sèche, que maintient du jonc ou de la ficelle, et
tantôt dans les pots. La première pratique ne convient
qu'à des plantes grasses, qui reprennent facilement après
avoir été privées de terre pendant un assez long-temps.
La seconde est bonne pour toute espèce de plantes lig-
neuses. La troisième semble pourtant mériter la préfé-
rence si l'emballage est fait avec de telles précautions, que
les pots ne puissent s'entrechoquer et se briser. Pour éviter
ce danger on les retient, ainsi que la terre qui les isole les
uns des autres, au moyen de petites traverses garnies de
mousse et fixées par les deux bouts à la paroi de la caisse.

" Ainsi disposées et abandonnées à elles-mêmes, les
plantes à l'abri de la sécheresse et de l'humidité, voyagent
pendant très long-temps, changeant de latitude et de cli-
mat, sans que leur santé soit sensiblement affectée. Elles
sont dans un état que l'on pourrait dire stationnaire.*

* One of the many misconceptions entertained respecting my cases is
that vegetation in them is arrested. A lady once called upon me, ima-
gining that I had invented a case in which half-blown Roses or other
flowers would remain *in statu quo* for an indefinite period. This opinion might be pardoned in one unacquainted with the laws of life, but I
certainly was not prepared to find it gravely stated by Mirbel. The
fact, of course, is that the plants, when thus supplied with all the means
of development, must either grow or die; and one of the causes of
failure in the travelling-case arises from the exuberant growth of some
one plant, which, by occupying all the surface of the glass, deprives
the others of light and life.

Il semble que chez elles la nutrition et la déperdition soient égales. La respiration continue ; les parties vertes conservent leur couleur, mais il n'y a point d'accroissement notable.

" Depuis plusieurs années, des envois faits de Londres à Calcutta, et de Calcutta à Londres, ont réussi au-delà de toute espérance. MM. Loddiges, frères, qui possèdent à Hackney, le plus riche jardin marchand qui soit en Europe, expédient sans cesse à la Nouvelle-Hollande, à la terre de Diemen, aux Indes-Orientales, des boîtes vides * qu'on leur expédie pleines. L'administration du Muséum d'Histoire Naturelle elle-même vient de recevoir pour la première fois, une de ces caisses dont elle est redevable à la bienveillance éclairée de M. Wallich, directeur du jardin de Calcutta. Cette caisse contenait quinze espèces précieuses, qui ne paraissent guère plus fatiguées que les plantes que nous retirons des serres au retour de la belle saison. Cependant la traversée avait été de huit à neuf mois. L'administration a renvoyé immédiatement à M. Wallich, en échange, dans une caisse faite sur la plan de la sienne, des végétaux de l'Europe australe, et des contrées chaudes de l'Amérique. A l'exemple du jardin du Roi, la famille Cels, dont le zèle héréditaire pour l'introduction en France des plantes exotiques est connu de tout le monde, a également adressé à M. Wallich, une caisse semblable remplie de végétaux.

" On ne saurait nier que l'usage des *serres de voyage*, qui, sans doute, sont encore susceptibles de modifications et de perfectionnements, ne doive contribuer beaucoup aux

---

* It is not customary with Messrs. Loddiges to send *empty* cases to their friends.

progrès de la phytologie ; et nous osons affirmer qu'il ne sera pas moins favorable à la naturalisation en Europe, d'une multitude d'espèces utiles ou agréables, qui compteraient déjà parmi les richesses de notre sol, si l'on avait trouvé plus tôt l'art de les y transporter vivantes.

" Nous souhaitons que des appareils semblables à ceux que nous avons décrits, soient mis à la disposition de MM les Médecins chargés spécialement de la récolte des objets d'Histoire Naturelle. La dépense est trop légère pour qu'elle soit un obstacle. Nous savons que l'administration du Muséum a fait parvenir à Toulon un petit modèle de *serre de voyage*, parfaitement exécuté, avec des instructions sur l'emploi de cet appareil. Le modèle et les instructions sont probablement à cette heure entre les mains de MM. les collecteurs."

———

(F)

Extract from the instructions given to Capt. Jas. Ross, *by the Royal Society of London.*

" Among those seeds which it is more particularly desirable to procure may be mentioned the arborescent *Compositæ* of St. Helena, and the native Coniferous plants of all countries, particularly the *Phyllocladus* or celery-leaved Pine, and the various species of *Athrotaxis* inhabiting the mountains of Van Diemen's Land. As the seeds of such plants are apt to suffer from long keeping, and as other instances may occur when it would be desirable to send home young plants instead of seeds, it would be advisable that the expedition should be supplied with *one* of Mr. Ward's glazed cases, to be used if occasion should arise."

———

(G)

Copy of a Letter from G. LODDIGES, Esq., to the Author.

Hackney,
February 18, 1842.

My Dear Sir,

In reply to your enquiries respecting the importation of living plants in your cases, I beg leave to say that my brother and I have, since 1835, made trial of more than 500 cases to and from various parts of the globe, with great variety of success; but have uniformly found, wherever your own directions were strictly attended to, — that is, when the cases were kept the whole voyage in the full exposure to the light, upon deck, and care taken to repair the glass immediately in cases of accident, — the plants have arrived in good condition; but we have never found this so well attended to as in those cases with which we have been favoured by your friends, and particularly by Capt. Mallard, of the Kinnear; indeed amongst all we have sent out or received, none have arrived in such good order as those brought by this gentleman. I wish we had more that possessed his love for Natural History, and would take the same care which he has done, as we should not then have to deplore the number of importations totally ruined, even in your cases, simply for the want of the light of day, and these too under the care of captains who engage that they shall be kept upon deck, when the moment we are out of sight they stow them away below, and they are never more thought of until their arrival: from experience in this mode of transportation we are enabled perfectly to see by their state whether they have been placed properly or not; for we find that there cannot be a worse mode of sending living plants, than in these same cases, so placed in the dark. Some of the cases have been opened in fine order after voyages of upwards of eight months : in short, nothing more appears to be wanting to ensure success in the importation of plants, than to place them in these boxes properly moistened, and to allow them the full benefit of light during the voyage.

I remain, My Dear Sir,

Ever your's most sincerely,

GEORGE LODDIGES.

To N. B. Ward, Esq.

———

(H)

Copy of a Letter from DR. STANGER, *to the Author.*

Wisbeach,
March 26, 1840.

My Dear Sir,

Permit me to give my testimony to the perfect security of your plan for the conveyance of plants in closed cases, having seen its benefits during a voyage from New South Wales to London round Cape Horn, in 1839.

On leaving Sydney I took on board two cases containing plants of the same kind; one closed, being covered with glass ; the other open, having a lid to be shut at pleasure.

The plants in the open case I nursed with the greatest care, keeping them below in cold and bad weather, and taking them up in fine. The only attention I paid to the other was to see it firmly lashed on deck, and when by accident a pane of glass was broken immediately replacing it with tin.

On my arrival in the Thames all the plants in the open case had perished except one, and that was looking very bad ; while those in the closed one were very healthy and vigorous, not one having in the least suffered ; in fact, they looked as well as the day they were put in, after a voyage of five months.

I may also add that the thermometer ranged during the voyage from 36° to 86° Fahr.

I remain, Dear Sir,

Your's truly,

W. STANGER.

To N. B. Ward, Esq.

———

(I)

Copy of a Letter from DR. LINDLEY, *to the Author.*

Hort. Soc.

January 15, 1842.

My Dear Sir,

As far as our experience goes your plant-cases are by far the best that have ever been contrived. We uniformly find the plants in them, even from India, in excellent order, provided the glass has not been broken, or they have not been over-watered when originally packed up. The latter arises from the packers not considering how little water is really requisite for plants which lose none of it. The former accident can hardly occur if the glass is well secured with a strong and close wireguard.

Pray believe me

Very truly yours,

JOHN LINDLEY.

To N. B. Ward, Esq.

(K)

Copy of a Letter from SIR W. J. HOOKER, *to the Author.*

Royal Botanic Garden, Kew,
January 24, 1842.

My Dear Sir,

I wish I could, from personal experience, add my testimony to that of Mr. Smith, respecting the importance of your cases for the transport of living plants from distant regions; but, during the short time that I have had the direction of the Royal Botanic Garden of Kew, no collections have arrived. From all, however, that I have seen of the healthy condition of plants soon after their arrival, that have been protected by your cases; and from all that I have heard from cultivators who have employed these cases to a considerable extent, I can have no hesitation in saying that your invention has been the means of introducing a great number of new and valuable plants to our gardens, from very distant countries, which would otherwise still have remained unknown to us.

Your cases for the in-door cultivation of tender plants are deservedly great favourites, and have contributed to the enjoyment of many families both in town and country. Splendid as is the hothouse and greenhouse collection at Woburn Abbey, I doubt whether that gives more pleasure to the noble proprietors and their numerous visitors than the beautiful little collection in " *Mr. Ward's case,*" that occupies a table in the library, and flourishes without requiring the skill of the gardener in its cultivation.

Pray believe me, My Dear Sir,

Very faithfully yours,

W. J. HOOKER,

DIRECTOR OF THE ROYAL BOTANIC GARDEN OF KEW.

To N. B. Ward, Esq.

(L)

Copy of a Letter from MR. J. SMITH, *to the Author.*

Royal Botanic Garden, Kew,
January 24, 1842.

Dear Sir,

In reply to your enquiry respecting the practical results obtained by adopting the plan of close-glazed cases, for the transfer of living plants from one country to another, I beg to say that the several cases which have arrived at this garden on that plan have shown that although all plants so treated may not succeed, still the deaths are but few in proportion to the number that we have witnessed in cases having open lattice or wire-work lids, covered with tarpauling or some such covering. It is much to be regretted that close-glazed cases were not in use during the years that the botanical collectors were employed in New Holland and the Cape of Good Hope, for this garden : a very great number of the plants which they sent home were always dead on their arrival, consequent on the imperfect protection during the voyage to this country ; therefore, from my experience, I have no hesitation in considering your plan the best for the purpose desired.

I am, Sir,
Your's truly,
J. SMITH.

To N. B. Ward, Esq.

(M)

Copy of a Letter from D. MOORE, Esq., *to the Author.*

Royal Botanic Garden, Glasnevin, Dublin,
February 1, 1842.

My Dear Sir,

I find all the species of ferns I have tried, to grow well either in glazed Wardian cases, under hand-lights, or in close frames, when the *external air can be excluded,* where some of the slender-growing kinds develope their fronds to such a degree of beauty and elegance as I have never observed excepting under such circumstances.

I may especially notice our rare and beautiful *Trichomanes speciosum,* Willd., which can be cultivated to very great perfection on this plan, and is here, at this time (1st February, 1842), in a fine state of fructification, producing larger fronds than it usually does in its native habitat. *Hymenophyllum Wilsoni,* Hook., and *H. Tunbridgense,* Sm., delight to grow in these close cases, and, when properly cultivated, attain to a larger size than they generally do in their habitats, producing fine fructiferous fronds.

Adiantum Capillus-Veneris, Linn., can only be seen to perfection in a cultivated state when grown in this manner, when it developes the fronds very large, and forms a beautiful object.

When the weather is very hot in summer I sometimes give them a sprinkling of water with the syringe, taking care to close the glasses as quick as possible, which greatly refreshes them, especially when in frames ; but during six or seven months of the year they never receive a drop of water artificially.

The various foreign species of *Lycopodia* I have tried in this way luxuriate amazingly. The only British species I have endeavoured to cultivate was *L. clavatum*, Linn., which grew very well, and when *hung up*, its long, slender, pendulous branches had a very graceful appearance.

I find many of the species of *Hepaticæ* thrive well in closed cases, especially those of the *Marchantiæ* and the larger species of *Jungermannia*, some of which have been cultivated here during the last three years, in a common frame, made as air-tight as possible.

The beautiful *Hygropila irrigua*, Taylor, grows well, and is now (1st February, 1842) in an incipient state of fructification.

Fegatella conica, Taylor, grows very strong, and also *Lunularia vulgaris*.

Jungermannia epiphylla, Linn., *furcata*, Linn., *asplenioides*, Linn., *emarginata*, Ehr., *nemorosa*, Linn., *Taylori*, Hook., *trilobatum*, Linn., *lævigatum*, Wils., *cochleariformis*, Weis, *tomentella*, Ehr., *Hutchinsiæ*, Hook., have all been successfully cultivated in this collection.

I remain, My Dear Sir,
Very truly your's,
D. MOORE.

To N. B. Ward, Esq.

(N)

Copy of a Letter from R. GRAHAM, Esq., *to the Author.*

Edinburgh,
February 4, 1842.

My Dear Sir,

I look upon your cases as important *practically* in reference to the introduction from abroad of plants transported with difficulty. In regard of the practical value of your observations I can speak positively, from the receipt at the Royal Botanic Garden here of boxes bringing in safety from New South Wales, from Cuba, from Jamaica, and from Bahia, plants which it might have been very difficult to import in any other way. Among these are some fine thriving plants of *Altingia excelsa*, several tree ferns, *Solandra daphnoides*, a species of *Malpighia*, and others. With a view to ascertain the practical value of the observations you have made, I have shut up in boxes, similar to those you recommend, a considerable number of plants, very various in constitution and organization; and I have been surprised to see with what perfect impunity some of these, which in general require very different treatment, bore the confinement in the damp atmosphere with which they were surrounded. Among these I may only mention Cactaceous plants, and different species of *Begonia* and *Cypripedium*. There are not many plants which flower freely so confined, but *Cypripedium venustum* thrives profusely, and flowers every year. Care should be taken not to make the earth wet when it is covered. It should be *damp*, but not *wet*.

Yours very truly,
ROBERT GRAHAM.

To N. B. Ward, Esq.

(O)

Copy of a Letter from MR. J. ANSELL, *to the Author.*

Hort. Society's Garden, Turnham Green,
March 2, 1842.

Sir,

In reply to your note of the 25th of February, I beg to give the following statement.

Four glazed boxes were taken out altogether, one of which was sent out before the others, and had been closed during eight months when I left it at the Niger. This box was filled with a mixture of half loam and half black mould, with the usual drainage at the bottom; and the plants, consisting of vines and figs, were put in in November, 1840, after which the box was left open, exposed to the open air till the frosts set in, when it was removed into a cool vinery. The box was closed on the 26th of January, 1841, after receiving a good watering, and the surface of the soil being covered with moss. It was then removed on board the vessel and placed on the deck, fully exposed to the sun. It remained on board till the middle of September, and I observed that the plants at that time were in a very healthy and growing condition. I cannot state anything concerning their further fate, as I was taken ill immediately after my arrival. The highest degree of heat to which they were exposed occurred when they were on shore at the confluence of the Niger and Chadda, where the thermometer was 90° in the shade.

I beg to observe that the old leaves were partly turned of a brown colour at the edges, probably scorched by the sun. As I went out myself in another vessel I had no means of protecting them from the sun, but when I received them at the mouth of the Niger they were partially shaded, and began to throw out young shoots.

The other boxes were likewise in very good condition, but the plants had not been confined for more than about five months.

I am, Sir,
Your obedient Servant,
JOHN ANSELL.

To N. B. Ward, Esq.

The Author will feel obliged by the results of any experiments in closed cases being communicated to him ; and, as he desires no other reward for his labours than the means of increasing his acquaintance with his favorite science, he will be thankful for any addition to his herbarium or general botanical collection.

Communications may be addressed to the Author at his residence, Wellclose Square, London; or to Mr. Pamplin, Botanical Agent, 9, Queen Street, Soho Square, London.